M000190007

ELECTRICAL SAFETY AND NFPA 70E®

2021 EDITION

Charles R. Miller

JONES & BARTLETT
L E A R N I N G

TABLE OF CONTENTS

TABLE OF CONTENTS (continued)

TABLE OF CONTENTS (continued)

TABLE OF CONTENTS (continued)

TABLE OF CONTENTS (continued)

# INTRODUCTION

The purpose of *Ugly's Electrical Safety and NFPA 70E®* is to provide recommendations in an abbreviated form. This pocket guide was created to provide a portable resource for field practitioners, electricians, maintenance personnel, contractors, inspectors, and engineers. This *Ugly's* guide is based on *NFPA 70E, Standard for Electrical Safety in the Workplace*. This publication is shorter than the actual standard and does not cover everything in the standard. This guide should be used with *NFPA 70E—not as a replacement for it*. For more complete information about electrical safety, see the following:

• *NFPA 70E—2021, Standard for Electrical Safety in the Workplace*

• *NFPA 70E: Handbook for Electrical Safety in the Workplace*

 **PROTECTIVE STRATEGIES**

The *NFPA 70E* standard describes safety-related work practices for workers who install, remove, inspect, operate, maintain, and disassemble electrical conductors and electric equipment, signaling and communications conductors and equipment, and raceways. The standard embraces five different protective strategies to eliminate or minimize exposure to electrical hazards. Requirements associated with the following five strategies discuss the most protective strategy to the least protective.

1. **Turn off the power.**

   Work de-energized, whenever possible (see electrically safe work condition). The standard recognizes that some work tasks, such as measuring voltage, require the circuit to be energized. When working within the limited approach boundary and the arc-flash boundary of exposed conductors and parts that are or might be energized, workers (or their supervisor) must use strategies 2–5.

2. **Perform a risk assessment.**

   Before work is started, use this procedure to determine the process to be used by the employee to identify hazards, assess risks, and implement risk control according to a hierarchy of methods. the hierarchy of preventive and protective risk control methods are elimination, substitution, engineering controls, awareness, administrative controls, and personal protective equipment (PPE). Elimination is the first (top) priority in the implementation of safety-related work practices, and PPE is the last priority **(see Figure 1)**. Risk assessment is an overall process that identifies hazards, estimates the likelihood of occurrence of injury or damage to health, estimates the potential severity of injury or damage to health, and determines whether protective measures are required.

2

PROTECTIVE STRATEGIES

FIGURE 1 **Hierarchy of controls.** (Courtesy of Charles R. Miller)

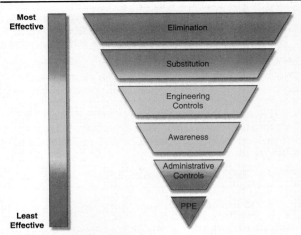

The hierarchy of preventive and protective risk control methods are elimination (first priority), substitution, engineering controls, awareness, administrative controls, and PPE (last priority).

3. **Use an energized electrical work permit.**
   Have the customer or employer sign an Energized Electrical Work Permit (EEWP). Additionally, if the worker is employed by the facility, the facility safety program should result in an EEWP.

4. **Plan the work and conduct a job briefing.**
   Before starting each job that involves exposure to electrical hazards, the employee in charge shall complete a job safety plan and conduct a job briefing with the employees involved.

⚡ QUALIFIED PERSONS (ELECTRICAL WORKERS)

5. Use personal protective equipment (PPE).

PPE includes, but is not limited to, arc-rated clothing, insulated tools, arc-rated face shields, arc-flash suits, safety glasses or safety goggles, hearing protection, and rubber insulating gloves with leather protectors

NFPA 70E defines a *qualified person* as "one who has demonstrated skills and knowledge related to the construction and operation of electrical equipment and installations and has received safety training to identify the hazards and reduce the associated risk." Only people who have received this training are permitted to do work within the limited approach boundary **(see Figure 2)**. Retraining in safety-related work practices and applicable changes in *NFPA 70E* shall be performed at intervals not to exceed 3 years.

FIGURE 2 Shock approach boundaries. (Courtesy of Charles R. Miller

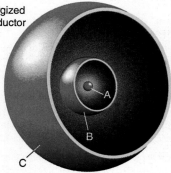

A. Exposed energized electrical conductor or circuit part

B. Restricted approach boundary

C. Limited approach boundary

The two types of boundaries associated with electric shock are limited approach boundary and restricted approach boundary. The limited approach boundary is the boundary farthest away from the energized electrical conductor or circuit part.

🔌 QUALIFIED PERSONS (ELECTRICAL WORKERS)

Qualified persons or workers who install and maintain electrical systems can include electricians, engineers, technicians, and building maintenance personnel who have received electrical training. Only qualified persons shall be permitted to work on electrical conductors or circuit parts that have not been put into an electrically safe work condition. It is up to the employer to determine who is qualified and meets their requirements.

A person is not necessarily qualified just because they hold an electrical license or have passed an electrical exam. A person who is trained and qualified on Vendor A switchgear, panelboards, and so on may not be qualified on Vendor B switchgear, panelboards, and so on, because the different brands of equipment may operate or disengage differently.

Protecting Others (That Is, Unqualified Persons)

NFPA 70E defines rules for protecting untrained workers from electrical hazards. Generally, people such as office workers; teachers; students; retail employees; customers; visitors; healthcare workers; cleaning crews; and workers in nonelectrical trades, such as plumbers, carpenters, and painters, are considered unqualified. As defined in Article 100, an unqualified person is a person who is not a qualified person.

NFPA 70E Requirements

Alerting Techniques

- Safety signs, barricades, and attendants keep unqualified persons away from places where an electrical hazard exists because of ongoing electrical work. Use safety signs, safety symbols, or accident-prevention tags where necessary to warn employees about electrical hazards that might endanger them. Such signs and tags shall meet the requirements of applicable state, federal, or local codes and standards.

QUALIFIED PERSONS (ELECTRICAL WORKERS)

- Use barricades in conjunction with safety signs where it is necessary to prevent or limit employee access to work areas containing energized conductors or circuit parts. Where the arc-flash boundary is greater than the limited approach boundary, barricades shall not b placed closer than the arc-flash boundary.

- If signs and barricades do not provide sufficient warning and protection from electrical hazards, assign an attendant to warn and protect employees. The attendant should manually signal and alert unqualified persons to stay outside the work area. As long as there i: a potential for employees to be exposed to the electrical hazards, the attendant shall remain in the area.

Look-Alike Equipment

- Where work will be performed on equipment that will be placed in an electrically safe condition and that equipment is in a work area with energized equipment that is similar in size, shape, and construction, one of the alerting methods in 130.7(E)(1), (2), or (3) shall be employed to prevent the employee from entering look-alike equipment. The alerting methods in 130.7(E)(1), (2), and (3) pertain to safety signs, barricades, and attendants.

Maintenance and Housekeeping

- Adequate maintenance and housekeeping help in making it possible to work safely on equipment by keeping the equipment in safe operating condition.

- Good maintenance includes technical factors such as ensuring that all electrical boxes have covers and all electrical equipment is adequately grounded.

- Good housekeeping includes nontechnical factors such as not using electrical rooms and closets for storage purposes.

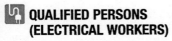

QUALIFIED PERSONS (ELECTRICAL WORKERS)

- Maintain electrical equipment in accordance with manufacturers' instructions or industry consensus standards to reduce the risk associated with failure. The equipment owner or the owner's designated representative shall be responsible for maintenance of the electrical equipment and documentation.

- The electrical safety program shall include elements that consider condition of maintenance of electrical equipment and systems.

- Working space required by other codes and standards shall not be used for storage. This space shall be kept clear to permit safe operation and maintenance of electrical equipment.

- Emergency access to switches and circuit breakers is essential. Space in front of electrical disconnecting means shall be kept clear.

- Employees shall not perform housekeeping duties inside the limited approach boundary, where there is a possibility of contact with energized electrical conductors or circuit parts, unless adequate safeguards (such as insulating equipment or barriers) are provided to prevent contact.

- Electrically conductive cleaning materials (including conductive solids such as steel wool, metalized cloth, and silicon carbide, as well as conductive liquid solutions) shall not be used inside the limited approach boundary unless procedures to prevent electrical contact are followed.

⚡ ELECTRICALLY SAFE WORK CONDITION

An "electrically safe work condition" is defined as a state in which an electrical conductor or circuit part has been disconnected from energized parts, locked/tagged in accordance with established standards, tested to verify the absence of voltage, and, if necessary, temporarily grounded for personnel protection. Establishing and verifying an electrically safe work condition shall include all of the following steps, which shall be performed in the order presented, if feasible:

1. Identify the power sources.

2. Disconnect power sources.

3. If possible, visually verify that power is disconnected.

4. Release stored electrical energy.

5. Block or relieve stored nonelectrical energy.

6. Apply lockout/tagout devices.

7. Test for the absence of voltage.

8. Install temporary protective grounding equipment if there is a possibility of induced voltages or stored electrical energy.

Warnings

- Only qualified persons are permitted to establish an electrically safe work condition.

- The process of establishing and verifying an electrically safe work condition is inherently hazardous. It requires contact (such as voltage testing) with exposed electrical conductors or circuit parts that might be energized. Electrical workers must wear appropriate PPE when performing some of the steps.

- Electrical conductors and equipment are considered energized until *an electrically safe work condition* is verified.

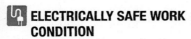

Eight-Step Process for Establishing and Verifying an Electrically Safe Work Condition

1. Identify the power source.
- Determine all possible sources of electric supply to the equipment to be worked on.
- Check electrical plans; one-line diagrams; panelboard schedules; and identifying tags, labels, and signs on electrical equipment.
- Most electrical equipment has a single source of supply, but sometimes equipment might have multiple sources. The multiple sources might include emergency generators, interactive power sources such as photovoltaic or fuel cell systems, and dual utility feeds for major industrial facilities.
- Sometimes, "illegal" circuits are installed that do not comply with *NEC* rules. These can create backfeed hazards after workers have disconnected all the electrical power sources they know about. An example is a 480v/120v lighting transformer. If the 480v is de-energized for working and temporary power is connected to the 120v panel, it must be verified that the 120v power cannot backfeed to the 480v system because 480v breakers were not opened and locked out.

2. Disconnect power sources.
- After properly interrupting the load current, open the disconnecting means for each source.
- Make sure the disconnecting means is capable of interrupting the load current.
- When the rating of a disconnect is not sufficient to interrupt the load current, the load must be removed by another operation before the handle is operated.
- Fuses are not considered disconnecting means, so a circuit cannot be de-energized merely by removing one or more plug or cartridge fuses. However, a pullout block (range fuse block) or safety switch with fuses is considered a disconnect. Operating the

◩ ELECTRICALLY SAFE WORK CONDITION

    switch or pulling out the fuse block disconnects all ungrounded (phase) conductors downstream of the circuit.

- On most premise wiring systems, only the ungrounded (phase) conductors are disconnected. The grounded (neutral) conductors normally are not interrupted intentionally.
- Attachment plugs for electrical appliances such as cooking and laundry equipment can be used as disconnects.

3. Visually verify that power is disconnected.

- Wherever possible, visually verify that all blades of the disconnecting means are fully open or that drawout-type circuit breakers are racked out to the test or fully disconnected position.
- Disconnecting means sometimes malfunction and fail to open all phase conductors when the handle is operated. After operating the disconnect's handle, a qualified person should open the equipment door or cover and look to see that a physical opening (air gap) exists in each blade of the disconnect.

4. Release stored electrical energy.

- Stored electric energy, which might endanger personnel, shall be released.
- Capacitors shall be discharged, and high-capacitance elements shall be short-circuited and grounded if the stored electric energy might endanger personnel.
- Discharge sources of stored energy such as capacitors used for power factor correction and motor starting.

     WARNING: Electrical workers must wear both shock and arc-flash PPE when performing this operation.

5. Block or relieve stored nonelectrical energy.

- Block or relieve stored nonelectrical energy in devices to the extent the circuit parts cannot be unintentionally energized by such devices.

⚡ ELECTRICALLY SAFE WORK CONDITION

- Examples of stored nonelectric energy include springs; elevated machine members; rotating flywheels; hydraulic systems; air, gas, steam, or water pressure; and so on.
- Dissipating or restraining methods might include repositioning, blocking, bleeding down, and so forth.

6. Apply lockout/tagout devices.

- Apply lockout/tagout devices in accordance with a documented and established procedure such as an employer's written electrical safety program. Normally, these devices are padlocks to keep the disconnecting means open (de-energized) and tags that identify the person(s) responsible for applying and removing the locks **(see Figure 3)**.

FIGURE 3   A lockout center. (Courtesy of Oberon Company)

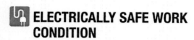

ELECTRICALLY SAFE WORK CONDITION

7. Test for the absence of voltage.

- Use an adequately rated, portable test instrument to test for the absence of voltage. An adequately rated, permanently mounted absence of voltage tester can be used to test for the absence of voltage of the conductors or circuit parts at the work location, provided it meets all of the following requirements: (1) It is permanently mounted and installed in accordance with the manufacturer's instructions and tests the conductors and circuit parts at the point of work; (2) it is listed and labeled for the purpose of testing for the absence of voltage; (3) it tests each phase conductor or circuit part both phase-to-phase and phase-to-ground; (4) the test device is verified as operating satisfactorily on any known voltage source before and after testing for the absence of voltage (**see Figure 4**).
- The functionality of the test instrument must be verified on any known voltage source both before and after using it to test for the absence of voltage.
- Use an adequately rated test instrument rated Category III or IV to test conductors and equipment operating at up to 600 or 1,000 volts.
- Test instruments rated Category II can be used on single-phase, 120-volt circuits.
- Test each phase conductor or circuit part to verify it is de-energized. Test each phase conductor or circuit part both phase-to-phase and phase-to-ground. Before and after each test, determine that the test instrument is operating satisfactorily. Noncontact test instruments can be used to test for the absence of voltage on electrical systems over 1,000 volts.
- The test shall also determine if any energized condition exists as a result of inadvertently induced voltage or unrelated voltage backfeed, even though specific parts of the circuit have been de-energized and are presumed to be safe.

> ⚠ *WARNING: Electrical workers must wear both shock and arc-flash PPE when performing this operation. Also, see the "Test Meter Safety Ratings" and the UL Category Ratings table on page 16.*

⚡ ELECTRICALLY SAFE WORK CONDITION

FIGURE 4 Permanently mounted absence of voltage (AVT) tester.
(Courtesy of Panduit Corp.)

8. **Install temporary protective grounding equipment if there is a possibility of induced voltages or stored electrical energy.**
   - Where the possibility of induced voltages or stored electrical energy exists, ground all circuit conductors and circuit parts before touching them.
   - Where it could be reasonably anticipated that the conductors or circuit parts being de-energized could contact other exposed energized conductors or circuit parts, apply temporary protective grounding equipment in accordance with the following:
     A. Temporary protective grounding equipment shall be placed at such locations and arranged in such a manner as to prevent each employee from being exposed to a shock hazard (i.e., hazardous differences in electrical potential). The location, sizing, and application of temporary protective grounding equipment shall be identified as part of the employer's job planning.

⚡ ELECTRICALLY SAFE WORK CONDITION

B. Temporary protective grounding equipment shall be capable of conducting the maximum fault current that could flow at the point of grounding for the time necessary to clear the fault.

C. Temporary protective grounding equipment and connections shall have an impedance low enough to cause immediate operation of protective devices in case of unintentional energizing of the electric conductors or circuit parts.

⚠️ *WARNING: Electrical workers must wear both shock and arc-flash PPE when performing this operation.*

After verification, the equipment and/or electrical source is now locked out (tagged out) and in an electrically safe work condition.

Electrically Safe Work Condition Established

After an electrically safe work condition has been established and verified, electrical energy has been removed from all conductors and equipment and cannot reappear unexpectedly. Under these circumstances, arc-rated and voltage-rated (insulated) electrical PPE is not needed, and unqualified persons can perform work such as cleaning and painting on or near electrical equipment. However, only electrically qualified persons should perform technical work within the scope of the *NEC*. This is true whether or not the electrical system is energized. When the electrical equipment is in an electrically safe work condition, it may not be permitted to remove all PPE. If the de-energized equipment is in an area where other PPE is normally required, these items shall not be removed. Examples of PPE normally required may include, but are not limited to, hearing protection, safety glasses, hard hats, and steel (or composite) toe work shoes or boots. An electrically safe work condition is not a procedure, it is a state wherein all hazardous electrical conductors or circuit parts to which a worker might be exposed are maintained in a de-energized state for the purpose of temporarily eliminating electrical hazards for the period of time for which the state is maintained.

TEST METER SAFETY RATINGS

Underwriters Laboratories, Inc. (UL), has four safety categories for test and measurement equipment. They are CAT I, CAT II, CAT III, and CAT IV. Note, these "CAT" ratings have no relationship to the four arc-flash PPE categories. CAT I offers the lowest level of protection, and CAT IV offers the highest level of protection.

- Meters are subject to transient voltage that can result from a lightning discharge, switching spike, or other occurrence.

- Voltmeters are assigned a transient rating based on the ability of the meter to continue working after experiencing a voltage spike.

- A worker must never connect any meter to electrical conductors or equipment with voltage or current higher than the rating of the meter itself.

- Transient overvoltages (spikes) caused by nearby lightning strikes, utility switching, motor starting, and capacitor switching can damage the electronic circuitry inside the meter, and can even cause it to explode while being used.

- Listed digital multimeters (DMMs) have internal fuses to protect the test instrument (and the person using it), but the worker must make sure that the meter is properly rated for the application before taking a reading **(see Table 1)**.

    ⚠ *WARNING: Only qualified persons shall perform tasks such as testing, troubleshooting, and voltage measuring on electrical equipment where an electrical hazard exists.*

TEST METER SAFETY RATINGS

TABLE 1   Meter Safety—UL Category Ratings

CAT I Isolated equipment	Equipment plugged into receptacle outlets and not directly connected (hard-wired) to the building supply system.
CAT II Directly connected circuits	Single-phase circuits supplied from panelboards. Equipment and appliances supplied by single-phase branch circuits.
CAT III Inside building circuits	Service equipment, panelboards, motor control centers. Three-phase feeders and branch circuits. Single-phase branch circuits supplied directly from the service equipment.
CAT IV Supply conductors Outdoor conductors	Service drop and service lateral conductors. Watt-hour meter. Line side of the main service disconnect. Outdoor feeders and branch circuits.

Location Is Everything

High-voltage transients can be caused by nearby lightning strikes, equipment within the facility, and normal utility switching operations. For this reason, choosing the right test meter for a particular measurement does not depend solely on circuit voltage or current, but on a combination of these factors and on the location of the item being tested in the premises wiring system **(see Figure 5)**.

- **CAT IV** meters can be used safely on outdoor conductors, the main lugs or main overcurrent (protection) device of service equipment, and watt-hour meters. CAT IV meters may be used in all category areas.

- **CAT III** meters can be used safely on power systems inside buildings and similar structures. These systems include panelboards, motor control centers, feeders, busways, motors, and hardwired luminaires (lighting fixtures). CAT III meters may also be used in CAT I and CAT II areas.

16

FIGURE 5   A meter typical of those used to test for voltage.
(Courtesy of Charles R. Miller)

- **CAT II** meters can be used safely on single-phase, receptacle-connected loads located more than 33 feet (10 meters) from a CAT III power source or more than 66 feet (20 meters) from a CAT IV source. CAT II meters may also be used in CAT I areas.

- **CAT I** meters are intended for use only on electronic equipment.

Meter Marking and Accessories

Meters are marked by the manufacturer with their category ratings. They are often dual-rated, such as CAT III-1000V/CAT IV-600V (see Figure 5). If a meter does not have a marked category rating, it is a CAT I device. Meters without a category marking provided by the manufacturer should not be used.

These energy safety ratings also apply to test leads, clamp-on adapters, and any other electrical accessories used with the meter. The lowest rating on any of these is the "weak link" that determines the overall category rating for that measurement **(see Figure 6)**.

FIGURE 6 Typical test lead that shows the category ratings.
(Courtesy of Charles R. Miller)

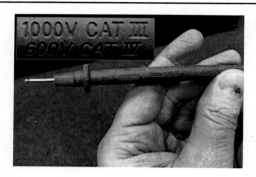

Select the Proper Rating

Select a meter with a rating that exceeds the anticipated application requirements for both the category and voltage. For most industrial work, 600V CAT III is the minimum acceptable.

Hazardous (Classified) Locations

The four UL category ratings do not apply to meters used in explosive atmospheres. Any meter used in a hazardous (classified) location as defined by the *NEC* must be "intrinsically safe."

Voltmeters and Test Instruments

Many different types of voltmeters and test instruments are available for purchase. In most instances, the devices work well within the

🔌 TEST METER SAFETY RATINGS

intended parameters. To avoid misapplication of the devices, workers must be familiar with the intended purpose of the device.

- To be safe from an electrical incident, the most important thing to know is whether an electrical conductor is energized.

- If the electrical conductor is energized, the next most important bit of information is the level of the voltage that is present.

- Only a voltmeter or test instrument can determine whether a conductor is energized. Although the voltage-detecting device is not worn or installed, as is the case with other PPE, the construction and integrity of the device are critical.

- Should a voltmeter or test instrument fail while in direct contact with an exposed energized conductor, an arcing fault may result.

- Should the voltmeter or test instrument fail to accurately sense the presence of voltage, electrocution may result. A worker can sense the presence of voltage by using an effective test instrument. A test instrument can also be used if it is tested on any known voltage source, used to check an item in question, and then used on any known voltage source again to make sure it is working.

- In most instances, the devices work well within the intended parameters. To avoid misapplication of the devices, be familiar with the intended purpose of the device.

Selection and Use

Select a voltmeter or test instrument that has a range commensurate with the expected circuit voltage.

- If the voltmeter or test instrument is intended to verify the absence of voltage, select a device intended for direct contact.

- If the level of voltage is not known, select a test instrument that has auto-ranging capability, but make sure the circuit voltage is lower than the maximum range of the test instrument.

 TEST METER SAFETY RATINGS

When a test instrument is being used, electrical injuries occur for one or more of the following reasons:

- The test instrument was misapplied or misused.

- The test instrument was selected improperly.

- The indication was misunderstood.

- The leads came out of the test instrument and touched the grounded enclosure, resulting in a short circuit or ground fault.

- The user's hand slipped off the end of the probe and contacted the energized conductor or circuit part.

- An internal failure occurred, and the test instrument exploded.

When the probes from a test instrument are in contact with energized conductors or circuit parts, the circuit under test experiences an additional circuit element. Current flows between one test instrument probe and the other.

- The amount of current depends on the internal impedance of the test instrument.

- Current is measured on a graduated scale (either analog or digital). In solenoid-type devices, the amount of current flow exerts a known magnetic force on the solenoid.

- The solenoid movement is graduated in voltage.

- For the test instrument to function effectively, the integrity of the current path through the measuring element is critical.

- The internal current path includes the probes, the plug in the case that accepts the probes, and internal components.

- In some cases, current flows through the switch used for changing scale. The switch could easily be set to the incorrect

🔧 TEST METER SAFETY RATINGS

position, which would destroy the meter and could cause an arcing fault.

- Devices with low internal impedance tend to discharge induced or static voltage. High internal impedance devices may measure all voltage, including induced and static voltage.

- It is important to understand the relationship of internal impedance to measuring voltage and the internal impedance of the current path.

- Some voltage-detecting devices do not require direct contact with an exposed energized conductor. Noncontact voltage-detecting devices sense the presence of an electrostatic or electromagnetic field. These devices look for inductive or capacitive coupling between the device and the conductor in question.

- Normally, noncontact voltage detectors provide an audible signal that voltage is present.

- Some devices also provide a visual signal. Depending on how the noncontact voltage detector works and the physical construction of the conductor in question, a null point may exist along the conductor's linear direction. These devices provide an initial indication of the presence of voltage, but they shall not be used to determine the absence of voltage.

- In accordance with *70E 120.5(7)*, an adequately rated portable test instrument shall be used to test each phase conductor or circuit part phase-to-phase and phase-to-ground. A noncontact voltage detector cannot be used to perform this type of test on electrical systems rated 1,000 volts or less. Noncontact test instruments can be used to test for the absence of voltage on electrical systems over 1,000 volts.

Duty Cycle

A duty cycle is intended to prevent the device from overheating. Several injuries occur every year because a solenoid-type device overheats and

explodes. Solenoid-type voltage-detecting devices are constructed by wrapping a small wire around a core to form the coil of a solenoid.

- Depending on the construction of the device, the manufacturer might assign a duty cycle to the device.

- The small wire usually has varnish for insulation. To avoid overheating the insulating varnish, a duty cycle is assigned to permit the coil to cool. A rule of thumb is that a solenoid voltage detector should be used for no more than 15 seconds without permitting the device to cool for at least another 15 seconds. The instructions from one manufacturer of solenoid-type voltage detectors, state for voltages above 240 volts, you must only connect to a voltage source for a maximum of 30 seconds and then disconnect for a minimum of 300 seconds.

- Manufacturers define the duty cycle on the label attached to the equipment.

Although a test instrument may be used to troubleshoot a circuit, the test instrument is a safety device.

- A test instrument provides crucial information for a worker to evaluate his or her exposure to an electrical hazard. The test instrument is PPE in the same vein as safety glasses or safety goggles.

- Electricians sometimes keep a voltmeter or test instrument in their toolbox with screwdrivers, socket wrenches, pipe wrenches, and so on. Workers then tend to view their voltmeter or test instrument as a hand tool that is used to troubleshoot a circuit, giving little thought to the fact that their lives depend on the integrity of their voltmeters or test instruments.

- Some workers carry voltmeters purchased at a hobby store because these devices may be cheaper. Still other workers carry a shirt pocket version of a noncontact voltage detector. These approaches may prevent workers from accurately sensing the presence or absence of voltage in a circuit, which is the first step in preventing injury.

TEST METER SAFETY RATINGS

UL 1244 **Requirements**

Purchase a test instrument that complies with specifications relevant for the intended use of the test instrument. The national consensus standard covering test instruments is *UL 1244, Standard for Electrical and Electronic Measuring and Testing Equipment.*

Requirements defined in this standard address issues that result in injuries associated with the construction of test instruments that were known when the standard was promulgated.

For instance, *UL 1244* requires that:

• The banana plugs that connect the leads to the test instrument be shielded to prevent contact with a grounded surface in case either of the plugs slips out of the receptacle.

• Probes contain a knurled section near the end to help prevent a worker's hand from slipping and contacting the energized conductor or circuit part.

• The test instrument be designed such that the mode selector switch cannot be a part of the active circuit.

• Adequately rated fuses, eliminating the risk of initiating an arcing fault from component failure, be present.

Only test instruments that are evaluated and comply with *UL 1244* may display the UL label. Unless products are marked as such, they do not comply with the standard's requirements.

Electrical systems are becoming increasingly complex. The amount of equipment that generates transient voltage spikes on a system is increasing.

• A transient voltage spike results from a static discharge in a lightning strike.

🔌 TEST METER SAFETY RATINGS

- Transients also might be the result of switching inductive loads.

- Normally, voltage spikes are a few microseconds in duration but can involve many hundreds of amperes. In some areas of North America, lightning discharges are common, especially in the spring and summer months.

- When a transient spike is in an electrical circuit, any equipment electrically connected to that circuit must be capable of handling the spike.

- Transients can destroy a test instrument that happens to be in contact with the electrical conductor simultaneously.

- Transients contain less energy as the distance from the source of the transient increases.

Static Discharge Categories

IEC 61010, Safety Requirements for Electrical Equipment for Measurement, Control, and Laboratory Use, and similar international standards establish a rating system for test instruments.

Test instruments are assigned to categories. The categories differentiate the ability of the devices to handle the energy in a transient condition (such as a spike in voltage caused by lightning or switching). The categories are established by the location in the circuit between the source of electricity and the point where the device will be used.

- Voltmeters and test instruments that can be used in the point of generation and transmission (where most energy is available) are Category (CAT) IV devices.

As the distance from the generator or transmission line to the point of use increases, the assigned category decreases.

🔌 TEST METER SAFETY RATINGS

- CAT III devices can be used safely with distribution level circuits such as motor control centers, load centers, and distribution panels.

- CAT II devices can be used safely with receptacles and utilization circuits. CAT I devices are intended for use with electronic equipment and circuits.

As the category decreases (from IV to I), the ability of the device to resist damage from transient overvoltage decreases. Therefore, CAT I devices are less likely to survive a transient overvoltage (spike).

In the international community, test instruments are assigned to categories according to their ability to function in various environments where transient currents are expected. The international standard that covers transient categories for test instruments is *IEC 61010, Safety Requirements for Electrical Equipment for Measurement, Control, and Laboratory Use.* The transient categories assigned by *IEC 61010* are CAT I, CAT II, CAT III, and CAT IV. These transient category ratings align with ratings established by the ANSI/ISA system. Test instrument ratings based on *ANSI/ISA S82.02.01* are equivalent to the ratings based on *IEC 61010.*

Purchase

Test instruments provide information that is critical to preventing injuries and are often kept in toolboxes that are subjected to physical damage. Test instruments that will be subjected to physical abuse should meet specifications that provide for protection. The authors recommend that the purchase order for test instruments require the device to have a UL label indicating compliance with *UL 1244, Electrical and Electronic Measuring and Testing Equipment.* The authors also recommend that test instruments for use in an industrial setting be assigned a CAT IV rating, as determined by *ANSI S82.02.01* or *IEC 61010.*

⚡ TEST METER SAFETY RATINGS

Inspection

Test instruments and equipment and all associated test leads, cables, power cords, probes, and connectors shall be visually inspected for external defects and damage before each use. Always make sure the test instrument is functioning normally before conducting the test, and then verify it is still functioning normally after testing for the absence of voltage. The inspection should verify the following information:

- The protective fuse is good.

- The case/enclosure is free from cracks and is not otherwise broken.

- The readout is clear and legible.

- The insulation on the leads is complete and undamaged.

- The shroud on all plugs is complete and undamaged.

- The finger guards are in place.

- The retractable probe covers are in place and functional.

- Each lead is continuous.

🔒 ENERGY CONTROL PROCEDURES

NFPA 70E defines two techniques for controlling hazardous electrical energy:

1. Simple lockout/tagout procedure

2. Complex lockout/tagout procedure

For more detailed information, see the following:

- Informative Annex G (in the back of *NFPA 70E*): Sample Lockout/Tagout Program

- Lockout/tagout procedures in *120.4* of *NFPA 70E*

Simple Lockout/Tagout Procedure

The simple lockout/tagout procedure involves qualified person(s) de-energizing one set of conductors or circuit part source for the sole purpose of safeguarding employees from exposure to electrical hazards **(see Figure 7)**:

- A written lockout/tagout plan for each application is not required.

- Each worker is responsible for their own lockout/tagout.

FIGURE 7 Simple lockout devices. (Courtesy of Larry W. Pace)

Complex Lockout/Tagout Procedure

Complex lockout/tagout is required under one or more of the following circumstances:

- Multiple energy sources and/or disconnects are involved.

- Multiple crews, crafts, or employers are involved.

- There are multiple locations.

- Particular sequences of de-energizing are needed.

- A job or task continues for more than one work period.

ENERGY CONTROL PROCEDURES

Power sources controlled by a complex lockout/tagout procedure
normally have more than one padlock and/or tag **(see Figure 8)**.

FIGURE 8   Complex lockout. (Courtesy of Charles R. Miller)

SAFETY PRACTICES WHEN ONE CONDUCTOR IS OR MIGHT BE ENERGIZED

Safe work practices for working within the limited approach boundary of energized or potentially energized electrical conductors or circuit parts consist of the following:

- Energized Electrical Work Permit

- Shock approach boundaries

- Arc-flash boundary

- PPE

- Arc-rated clothing

SAFE ELECTRICAL WORK PRACTICES

Equipment Labeling

Electrical equipment, such as switchboards, panelboards, industrial control panels, meter socket enclosures, and motor control centers that are in other than dwelling units and are likely to require examination, adjustment, servicing, or maintenance while energized, shall be field marked with a label containing the following information:

1. **Nominal system voltage**

2. **Arc-flash boundary**

3. **At least one of the following:**
   - Available incident energy and the corresponding working distance **(see Figure 9)**, or the arc-flash PPE category in Table 130.7(C)(15)(a) or Table 130.7(C)(15)(b) for the equipment **(see Figure 10)**, but not both
   - Minimum arc rating of clothing
   - Site-specific level of PPE

Exception No. 1: Unless changes in electrical distribution system(s) render the label inaccurate, labels applied prior to the effective date of the 2021 edition of 70E shall be acceptable if they complied with the requirements for equipment labeling in the standard in effect at the time the labels were applied.

Exception No. 2: In supervised industrial installations where conditions of maintenance and engineering supervision ensure that only qualified persons monitor and service the system, the information required in 130.5(H)(1) through 130.5(H)(3) shall be permitted to be documented in a manner that is readily available to persons likely to perform examination, servicing, maintenance, and operation of the equipment while energized.

The method of calculating, as well as data to support the information for the label, shall be documented. The data shall be reviewed for accuracy at intervals not to exceed 5 years. Where the review of the data identifies a change that renders the label inaccurate, the label shall be updated.

The owner of the electrical equipment shall be responsible for the documentation, installation, and maintenance of the field-marked label.

FIGURE 9   Equipment label with available incident energy.
(Courtesy of Charles R. Miller)

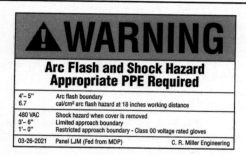

FIGURE 10   Equipment label with arc-flash PPE category.
(Courtesy of Charles R. Miller)

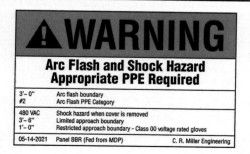

SAFE ELECTRICAL WORK PRACTICES

An arc-flash hazard calculation study is just one of several methods that can be used to find the information required on an equipment warning label. An arc-flash hazard calculation study is used to find information such as the following:

- The incident energy exposure level based on the working distance of the employee's face and chest areas from a prospective arc source

- The arc-flash boundary

- Arc-rated clothing and other PPE used by the worker, based on the incident energy exposure level

- Shock protection boundaries

Electrical Hazards

Sometimes it is not feasible to establish an *electrically safe work condition* (turn off the power) before doing repair or maintenance work on electrical systems. For instance, taking a current reading with a clamp-on ammeter requires the circuit to be energized. When working on exposed conductors and equipment operating at 50 to 600 volts, *NFPA 70E* requires the following:

- A risk assessment. As used in *70E*, arc-flash risk assessment and shock risk assessment are types of risk assessments.

- An Energized Electrical Work Permit.

- A shock risk assessment to determine the voltage to which personnel will be exposed, the shock boundary requirements, and the PPE necessary in order to minimize the possibility of electric shock to personnel.

- An arc-flash risk assessment to determine whether an arc-flash hazard exists. If an arc-flash hazard exists, the risk assessment shall determine appropriate safety-related work practices, the arc flash boundary, and the PPE to be used within the arc flash boundary. Before starting each job that involves exposure to electrical hazards, the employee in charge shall complete a job safety plan and conduct a job briefing with the employees involved.

SAFE ELECTRICAL WORK PRACTICES

Other safety precautions include the following:

- Be alert at all times when working within the limited approach boundary of energized electrical conductors or circuit parts operating voltages equal to or greater than 50 volts, or in work situations where other electrical hazards exist.

- Do not reach blindly into areas that might contain exposed energized electrical conductors or circuit parts where an electrical hazard exists.

- Make sure there is adequate illumination for energized work.

- Do not perform energized work if the view is obstructed.

- Do not wear conductive items such as jewelry, watchbands, metal-frame glasses, clothing with metallic threads, or metallic body piercings. Steel-toed work boots required by Occupational Safety and Health Administration (OSHA) regulations are acceptable because leather or fabric covers the metal.

Energized Electrical Work Permit

When Is a Work Permit Required?

When energized work is performed as permitted in accordance with 110.4, an Energized Electrical Work Permit shall be required and documented under any of the following conditions:

(1) When work is performed within the restricted approach boundary

(2) When the employee interacts with the equipment when conductors or circuit parts are not exposed, but an increased likelihood of injury from an exposure to an arc-flash hazard exists.

What Information Is Required on a Work Permit?

The person authorizing the work (management, safety officer, or owner) must sign an Energized Electrical Work Permit like the one

⬛ SAFE ELECTRICAL WORK PRACTICES

shown in *NFPA 70E-2021*, Annex J, Figure J.1. This permit contains the following:

- Description of the circuit and equipment to be worked on and their location

- Description of the work to be performed

- Justification of why the work must be performed with the circuit/equipment energized (see the discussion following this list)

- Description of the safe work practices to be employed

- Results of the shock risk assessment, which shall include the following:

  1. Voltage to which personnel will be exposed

  2. Limited approach boundary

  3. Restricted approach boundary

  4. Personal and other protective equipment required by *NFPA 70E* to safely perform the assigned task and to protect against the shock hazard

- Results of the arc-flash risk assessment, which shall include the following:

  1. Available incident energy at the working distance or arc-flash PPE category

  2. Personal and other protective equipment required by *NFPA 70E* to protect against the arc-flash hazard

  3. Arc-flash boundary

- Description of methods to be used to keep unqualified persons out of the work area

🔌 SAFE ELECTRICAL WORK PRACTICES

- Evidence of completion of a job briefing, including a discussion of any job-specific hazards

- The signature(s) of the person(s) authorizing the work to be performed with the circuit/equipment energized

What Is Normal Operating Condition of Electric Equipment?

Normal operating conditions of electric equipment shall be permitted where a normal operating condition exists. A normal operating condition exists when all of the following conditions are satisfied:

1. The equipment is properly installed.

2. The equipment is properly maintained.

3. The equipment is used in accordance with instructions included in the listing and labeling and in accordance with manufacturer's instructions.

4. The equipment doors are closed and secured.

5. All equipment covers are in place and secured.

6. There is no evidence of impending failure.

As used in *70E*, the term "properly installed" means the equipment is installed in accordance with the manufacturer's recommendations as well as applicable industry codes and standards. The term "properly maintained" means the equipment has been maintained in accordance with the manufacturer's recommendations as well as applicable industry codes and standards. The term "evidence of impending failure" means there is evidence such as arcing, overheating, loose or bound equipment parts, visible damage, or deterioration.

SAFE ELECTRICAL WORK PRACTICES

Reasons That Justify Working with the Circuit/Equipment Energized

Additional Hazards or Increased Risk

NFPA 70E allows energized work to be performed if de-energizing introduces additional hazards or increased risk. Examples of such hazards include, but are not limited to, interrupting life-support equipment, deactivating emergency alarm systems, and shutting down ventilation equipment in hazardous (classified) locations.

Design or Operational Limitations

NFPA 70E allows energized work to be performed if de-energizing is infeasible due to equipment design or operational limitations.

Examples of such infeasibility include performing diagnostics and testing (e.g., start-up or troubleshooting) of electric circuits that can only be performed with the circuit energized and work on circuits that form an integral part of a continuous process that would otherwise need to be completely shut down in order to permit work on one circuit or piece of equipment.

The Difference Between *INCONVENIENT* and *INFEASIBLE*

The Task

Ballasts in fluorescent ceiling fixtures in an office need to be replaced. The customer does not want to turn off the branch circuit(s) feeding the fixtures during normal working hours for fear of inconveniencing workers and reducing productivity.

The Best Solution

The best solution is to replace the ballasts at night or over a weekend, when an *electrically safe work condition* can be established and electricians can work using only minimal PPE (eye protection and work gloves).

 SAFE ELECTRICAL WORK PRACTICES

REMEMBER: Turning off the power is always the safest way to work!

Acceptable Alternate Solution

Electrical workers can have the customer complete an Energized Electrical Work Permit. Note that the work must be justified as defined by *NFPA 70E*. The workers can then do the work while the circuit is energized, during normal office hours, with the electricians using PPE and following all other safety precautions required by *NFPA 70E*.

This is an *acceptable* solution because it complies with *NFPA 70E* and provides a degree of protection for workers. However, turning off the power is always the safest way to work on electrical conductors and equipment.

NOTE: Working energized is a last resort and always requires PPE.

Shock Protection

Determine Shock Approach Boundaries

The second safety-related work practice when working with the circuit/equipment energized is determining the shock approach boundaries; this is referred to as a shock risk assessment. A shock risk assessment shall be performed:

- To identify shock hazards

- To estimate the likelihood of occurrence of injury or damage to health and the potential severity of injury or damage to health

- To determine if additional protective measures are required, including the use of PPE

If additional protective measures are required, they shall be selected and implemented according to the hierarchy of risk control identified in 110.5(H)(3). When the additional protective measures include the use of PPE, the following shall be determined:

- The voltage to which personnel will be exposed

38

⚡ SAFE ELECTRICAL WORK PRACTICES

- The boundary requirements

- The personal and other protective equipment required by this standard to protect against the shock hazard.

These boundaries help protect against shock and electrocution.

Approach boundaries are identified as limited approach boundary and restricted approach boundary **(see Figure 11)**. Crossing one of these approach boundaries increases the chance that a worker might contact an exposed energized electrical conductor or circuit part.

FIGURE 11   Shock approach boundaries. (Courtesy of Charles R. Miller)

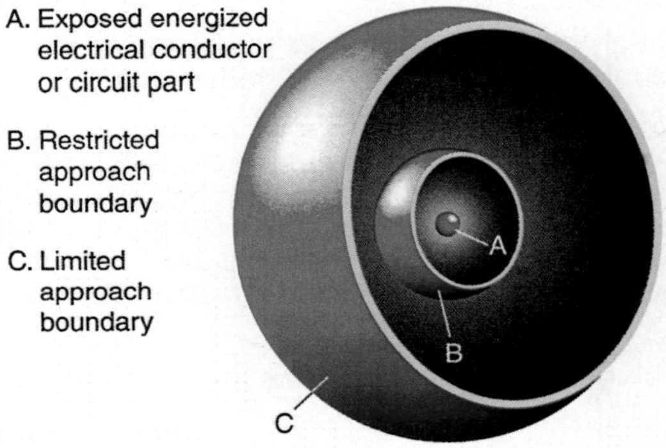

A. Exposed energized electrical conductor or circuit part

B. Restricted approach boundary

C. Limited approach boundary

The two types of boundaries associated with electrical shock are limited approach boundary and restricted approach boundary. The shock protection boundaries and the arc-flash boundary are independent of each other. Note, the arc-flash boundary is not shown in this illustration.

SAFE ELECTRICAL WORK PRACTICES

Where approaching personnel are exposed to energized electrical conductors or circuit parts, the approach boundaries are as follows:

- *Limited approach boundary*—This boundary is an approach limit at a distance from an exposed energized electrical conductor or circuit part within which a shock hazard exists. This boundary is larger for movable conductors than for fixed circuit parts.

- *Restricted approach boundary*—This boundary is an approach limit at a distance from an exposed energized electrical conductor or circuit part within which there is an increased likelihood of electric shock, due to electrical arc-over combined with inadvertent movement. It allows for the fact that a person's hand or tool might slip, or someone else might jostle the worker from behind **(see Table 2A and Table 2B)**.

Limited Approach Boundary

- *Qualified persons.* Only electrically qualified persons can work inside a limited approach boundary without special precautions.

- *Unqualified persons.* Persons such as management employees, cleaning crews, painters, and other construction trades can cross the limited approach boundary only when continuously escorted and advised of the possible hazards by an electrically qualified person. Otherwise, unqualified persons must be kept outside the limited approach boundary by barricades, warning signs, or attendants.

- When electrical conductors and circuit parts are in an electrically safe work condition, the limited and restricted approach boundaries do not exist.

SAFE ELECTRICAL WORK PRACTICES

TABLE 2A 130.4(E)(a) Approach Boundaries for Shock Protection, Alternating-Current Systems

Phase-to-Phase Voltage	Limited Approach Boundary Movable	Limited Approach Boundary Fixed	Restricted Approach Boundary
Less than 50	Not specified	Not specified	Not specified
50–150*	10 ft	3 ft 6 in.	Avoid contact
151–750	10 ft	3 ft 6 in.	1 ft
751–15,000	10 ft	5 ft	2 ft 2 in.
15,001–36,000	10 ft	6 ft	2 ft 9 in.
36,001–46,000	10 ft	8 ft	2 ft 9 in.
46,001–72,500	10 ft	8 ft	3 ft 6 in.
72,600–121,000	10 ft 8 in.	8 ft	3 ft 6 in.

*This includes circuits where the exposure does not exceed 120 volts nominal.

TABLE 2B 130.4(E)(b) Approach Boundaries for Shock Protection, Direct-Current System

Nominal Potential Difference	Limited Approach Boundary Movable	Limited Approach Boundary Fixed	Restricted Approach Boundary
Less than 50	Not specified	Not specified	Not specified
50 to 300	10 ft	3 ft 6 in.	Avoid contact
301 to 1,000	10 ft	3 ft 6 in.	1 ft
1,100 to 5,000	10 ft	5 ft	1 ft 5 in.
5,000 to 15,000	10 ft	5 ft	2 ft 2 in.
15,100 to 45,000	10 ft	8 ft	2 ft 9 in.

⚡ SAFE ELECTRICAL WORK PRACTICES

Restricted Approach Boundary

- *Qualified persons.* Electrically qualified persons can cross a restricted approach boundary only when wearing and using PPE, such as adequately rated gloves and tools, for protection from shock.

- *Unqualified persons.* Under no circumstances are unqualified persons permitted to cross the restricted approach boundary.

Not Arc-Flash Protection!

Limited and restricted approach boundaries are for shock protection only. They have nothing to do with protecting workers from arc-flash and arc-blast hazards, or selecting arc-rated clothing. See the following section.

🔌 SHOCK PROTECTION

Insulating Rubber Gloves

- Workers shall wear rubber insulating gloves with leather protectors where there is a danger of hand injury from electric shock due to contact with energized electrical conductors or circuit parts.

- Insulating rubber gloves increase the resistance of the current path between an energized electrical conductor and a person's skin.

- By increasing the impedance, the gloves reduce the amount of current that flows through the person's body as a result of contact (accidental or intentional).

- The amount of current that flows through body tissue is directly proportional to the kind and degree of injury.

If the insulation value of a worker's glove is overcome by the voltage potential of the electrical conductor, the glove resistance decreases to the point that dangerous currents might flow through the worker's body. Workers must have the ability to select an insulating glove that cannot be affected by the voltage potential of the circuit.

NOTE: Heavy-duty leather gloves or arc-rated gloves shall be worn where required for arc-flash protection. If rubber insulating gloves with leather protectors are used for shock protection, additional leather or arc-rated gloves are not required. Rubber insulating gloves with leather protectors provide arc flash protection in addition to shock protection. The combination of rubber insulating gloves with leather protectors satisfies the arc-flash protection requirement.

Rubber insulating gloves shall be permitted to be used without leather protectors under the following conditions:

1. There shall be no activity performed that risks cutting or damaging the glove.

2. The rubber insulating gloves shall be electrically retested before reuse.

🔌 SHOCK PROTECTION

3. The voltage rating of the rubber insulating gloves shall be reduced by 50% for class 00 and by one whole class for classes 0 through 4.

Hazards

- The rubber used in the insulating layer can burn, but it is difficult to ignite.

- The leather protectors provide additional protection for the hands against arc-flash exposure.

- Heavy-duty leather gloves are made entirely of leather with minimum thickness of 0.03 in. (0.7 mm). They are unlined or lined with nonflammable, nonmelting fabrics. Heavy-duty leather gloves have been shown to have ATPV values in excess of 10 cal/cm^2 **(see Figure 12)**.

FIGURE 12   Gloves. (Courtesy of Oberon Company)

SHOCK PROTECTION

Size and Style

Voltage-rated gloves should fit the hands of the person wearing them. Glove size for a specific hand is determined by measuring the circumference of the palm at its widest point. The palm measurement is the size that should be ordered **(see Figure 13)**.

FIGURE 13 Measuring the hand for glove fit. (Courtesy of Charles R. Miller)

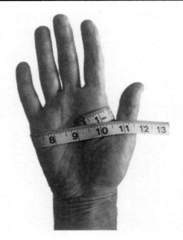

Leather Protectors

- Where insulating rubber gloves are used for shock protection, leather protectors shall be worn over the rubber gloves.

- The function of leather protectors, sometimes called "leathers," is to reduce the chance of damage to the insulating glove.

45

🔌 SHOCK PROTECTION

- When a worker touches an energized conductor with the gloves, the entire leather protector is energized at the same voltage of the conductor.

- The leather protector must not contact any uninsulated body part.

- Leather is conductive. Therefore, the insulating rubber glove must extend beyond the leather protector by a length that is sufficient to eliminate the chance of creepage between the leather protector and the worker's skin.

- The distance between the end of the cuff of the protector and the end of the cuff of the rubber glove shall not be less than 1/2 inch for glove Class 00 and Class 0, 1 inch for Class 1, 2 inches for Class 2, 3 inches for Class 3, and 4 inches for Class 4 (see ASTM F496) **(see Figure 14)**.

FIGURE 14   Minimum gap distances. (Courtesy of Charles R. Miller)

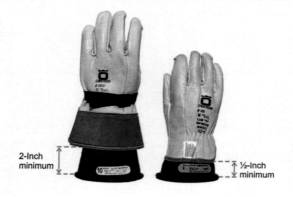

2-Inch minimum

½-Inch minimum

FIGURE 15 **Leather protectors.** (Courtesy of Oberon Company)

- Gloves insulate the hands of the worker from current flow (shock) because the rubber is an insulator.

- Leather is not an insulator; therefore, leather protectors protect only the gloves, not the hands. In the event of an arc flash, leather protectors become the protecting component **(see Figure 15)**.

NOTE: Because leather is conductive, do not use leather protectors alone for protection against electric shock.

Selection and Use of Voltage-Rated Gloves

- Rubber gloves used for shock protection should be rated gloves or voltage-rated (insulating) gloves.

- Although many different types of rubber gloves are on the market, unless they meet consensus requirements as voltage-rated (insulating) gloves, they should never be used for shock protection **(see Figure 16)**.

FIGURE 16   Voltage-rated gloves. (Courtesy of Charles R. Miller)

⚡ SHOCK PROTECTION

Maintenance and Inspection

- Electrical protective equipment shall be maintained in a safe, reliable condition.

- Gloves and leather protectors must be inspected before each use **(see Figure 17)**. For a flowchart for glove inspection, **see Figure 19**.

- Insulating equipment shall be inspected for damage before each day's use and immediately following any incident that can reasonably be suspected of having caused damage.

FIGURE 17   Damaged gloves.

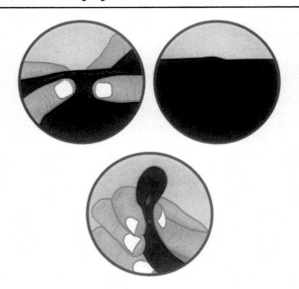

SHOCK PROTECTION

FIGURE 18   **Glove inspection.** (Courtesy of Charles R. Miller)

- Insulating gloves shall be given an air test along with the inspection **(see Figure 18)**.

- Nationally recognized standards require visual examination and inspection of both the insulating glove and the leather protector before use; the reason for inspecting them is to prevent shock and electrocution.

- Any damage to either the insulating material or the leather protectors can result in the wearer being shocked or electrocuted when direct contact is made with an exposed energized conductor.

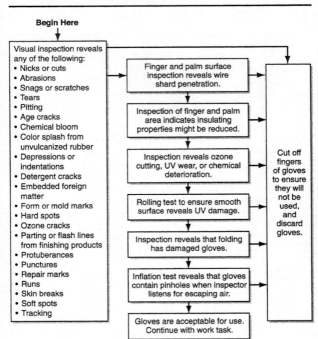 SHOCK PROTECTION

Sleeves

- Workers shall wear rubber insulating gloves with leather protectors and rubber insulating sleeves where there is a danger of hand and arm injury from electric shock due to contact with energized electrical conductors or circuit parts.

FIGURE 19 Glove inspection flowchart.

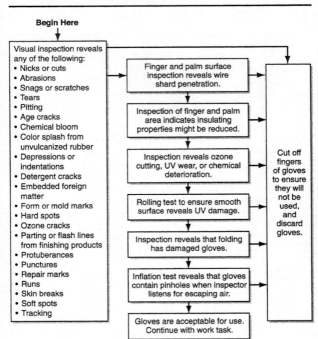

SHOCK PROTECTION

FIGURE 20 Worker wearing sleeves correctly. (© Ariel Skelley/
DigitalVision/Getty Images)

- Sleeves serve to increase the impedance of the current path between
  an exposed energized conductor and a person attempting to
  manipulate the conductor or another circuit part **(see Figure 20)**.

- By increasing the impedance, the sleeves reduce the amount of
  current flow to a predetermined level.

- The objective is to reduce the current flow to a level that cannot harm
  the person wearing the sleeves. Sleeves must never be worn without
  voltage-rated gloves.

- Sleeves are held in place by buttons and straps or a harness **(see
  Figure 21)**.

⌁ SHOCK PROTECTION

FIGURE 21 Straps and harness used to hold sleeves in place.
(Courtesy of Salisbury Electrical Safety, LLC)

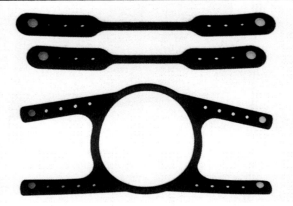

Mats and Matting

- A rubber mat is essentially a subset of matting **(see Figure 22)**.

- Mats are intended to be installed on the floor in front of electrical equipment.

How Mats and Matting Provide Protection

- Muscle tissue reacts to current flow through the body. Shock and electrocution are the result of current flow.

- Voltage acts as the pressure that forces the current to flow. As the amount of current increases, the body reacts more violently.

- If the current flow can be limited to a value that has no deleterious effect, neither shock nor electrocution are possible.

FIGURE 22   Matting. (Courtesy of Salisbury Electrical Safety, LLC)

- The intended role of mats and matting is to reduce the amount of current flow should contact with an exposed energized conductor or circuit part be made.

If a worker touches an exposed energized conductor (and no other conductor or grounded surface) while standing on an appropriately rated mat, the amount of current flow through the worker's body will be small, and the worker will not be injured.

Rubber gloves and blankets insert insulation between a worker and an energized component. The worker can be in contact with earth ground, and still current cannot flow because the gloves or blanket reduce the chance of contact with an energized component **(see Figure 23)**.

Blankets

In normal use, blankets come into direct contact with the electrical conductor. Consequently, the prohibited approach boundary is penetrated each time a blanket is installed.

FIGURE 23   Rubber gloves and mats add resistance between worker and earth ground. (Courtesy of Salisbury Electrical Safety, LLC)

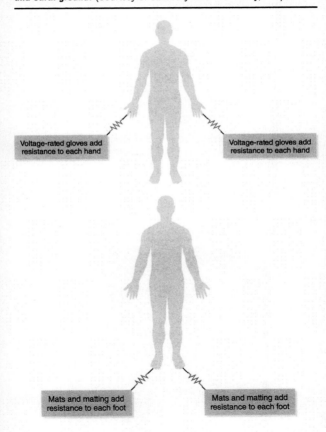

Voltage-rated gloves add resistance to each hand

Voltage-rated gloves add resistance to each hand

Mats and matting add resistance to each foot

Mats and matting add resistance to each foot

⚡ SHOCK PROTECTION

- Current consensus standards require each employer to determine necessary procedural and administrative actions when the restricted approach boundary is penetrated.

- The circuit should be de-energized before the blanket(s) is (are) installed.

- The worker should ensure that the circuit is clear of any potential fault prior to re-energizing the circuit.

Sometimes it is not possible to de-energize the circuit to install blankets for an emergency work task. In that instance, strict adherence to procedural requirements is critical.

- Blankets normally are wrapped around an electrical conductor in such a way that the worker is less likely to contact the conductor when executing the work task **(see Figure 24)**.

- The blankets may be held in place by nonconductive cable ties, buttons, or clamp pins. Electricians frequently refer to the clamp pins

FIGURE 24 Blankets.

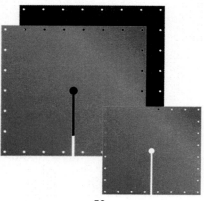

as clothespins, because they have the general appearance of large clothespins. Blankets with hook-and-pile (Velcro®) fastening along the edges are also available.

• Blankets are intended to provide temporary insulation on an electrical conductor and should not be left in place after the work task is complete.

Hazards

Insulating blankets are ineffective as protection from other electrical hazards such as arc flash or arc blast. Rubber blankets are held in place by clothespins, cable ties, or other means, and the device used to hold the blanket in place could become a missile should an arcing fault occur, thus introducing a new hazard.

An arcing fault is electrical current flowing in air between two conductors.

• Arcing faults may be the result of component failure. However, a worker making inadvertent contact with an energized conductor usually initiates an arcing fault.

• An adequately rated blanket reduces the chance of initiating a fault on the conductor. However, the blanket will have no effect in reducing a worker's exposure to the thermal and blast effects of an arcing fault.

An arcing fault also produces a significant pressure wave. Fasteners could become projectiles should an arcing fault occur. Placement of fasteners is an important consideration when positioning the fasteners on the temporary blanket. Buttons are more likely to cause injury than cable ties **(see Figures 25 and 26)**.

FIGURE 25 Buttons hold rubber blankets in place.

FIGURE 26 Clothespins hold rubber blankets in place.

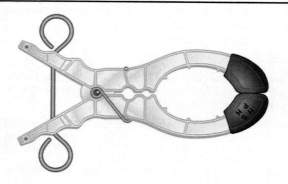

Footwear

For current to flow through a worker's body, the worker must contact an exposed energized conductor and another conductor, such as earth ground. If the worker wears footwear that is adequately rated, foot contact with earth ground is reduced or eliminated.

Dielectric footwear must be constructed as defined in the standard *ASTM F2413, Standard Specification for Performance Requirements for Protective (Safety) Toe Cap Footwear* **(see Figure 27)**. Tests to

FIGURE 27 Dielectric footwear.

determine the electrical protective nature of dielectric footwear are defined in *ASTM F1116, Standard Test Method for Determining Dielectric Strength of Dielectric Footwear*. These national consensus standards provide assurance that the footwear meets normal expectations for protecting a worker's feet from both physical and electrical shock hazard.

Leather footwear or dielectric footwear or both provide some arc flash protection to the feet and shall be used in all exposures greater than 4 cal/cm^2 (16.75 J/cm^2). Footwear other than leather or dielectric shall be permitted to be used provided it has been tested to demonstrate no ignition, melting, or dripping at the estimated incident energy exposure or the minimum arc rating for the respective arc flash PPE category.

Live-Line Tools (Hot sticks)

- Electricians have used the term "hot stick" for many years. The term suggests that the device is intended to contact an energized conductor.

- When an electrician says that a conductor is "hot," they probably are talking about the fact that a conductor is energized relative to earth ground. However, the electrician could be talking about the thermal condition of a conductor that is overloaded.

🔌 SHOCK PROTECTION

- In another example, when thermographers perform analyses on electrical conductors and terminations, the term "hot" refers to the thermal condition of the conductor or termination.

Use of the term "hot" to mean both "energized" and "high thermal temperature" can be very confusing. The term "hot stick," then, is a misnomer, although it is the most commonly used term.

- OSHA regulations use the term "hot" to refer to a thermal condition instead of an energized condition.

- In OSHA regulations, the term "hot stick" is used only in an appendix, and then only in an attempt to ensure complete understanding by the user. In *29 CFR 1910.269(j)*, OSHA uses the descriptive term "live-line" tools instead of "hot stick" to discuss tools constructed from fiberglass-reinforced plastic (FRP).

- In common usage, both "hot stick" and "live-line tools" refer to a family of tools constructed from FRP material that are used to perform specific actions on an energized conductor or component **(see Figure 28)**.

Hazards

- Should an arcing fault occur when the worker is using a live-line tool, the worker could be exposed to the thermal hazard associated with the arcing fault. The live-line tool provides no protection from this hazard.

FIGURE 28   Live-line tool (hot stick). (Courtesy of Oberon Company)

🔌 SHOCK PROTECTION

- The chance of initiating an arcing fault by contacting an energized component with a live-line tool is remote. However, equipment or devices that may be attached to the end of a live-line tool can initiate an arcing fault **(see Figure 29)**.

FIGURE 29 Arcing fault on an overhead line. (Courtesy of D. Ray Crow)

SHOCK PROTECTION

- The length of the live-line tool is the critical component of preventing thermal injury to an otherwise unprotected worker.

Selection and Use

- Customers expect that utilities will keep electrical energy available for their use in homes, businesses, commercial endeavors, and industrial locations. Utilities and manufacturers respond to that demand by developing equipment, tools, and work processes that enable maximum system availability for customers.

- Customers also expect that utilities will provide the electricity at minimum costs. Utilities respond by developing equipment that minimizes the cost of the installation to the consumer while maintaining the integrity of the service. For instance, fuse clips become disconnecting devices. Devices that permit an appropriately assembled live-line tool to operate the disconnecting device replace ground-operated switching devices.

- Frequently, the means to disconnect the utility will be located at the top of a pole (see Figure 30). A live-line tool enables a worker to operate the disconnecting means from the ground or from a bucket truck.

Disconnecting means that are operated by a live-line tool cost much less than gang-operated switches (see Figure 31). Gang-operated means that all phases are connected together mechanically so that all phases break and make together.

- Generally, larger manufacturing facilities require immediate access to disconnecting means. These facilities normally install gang-operated switches, enclosed fusible switches, or circuit breakers in an enclosure.

FIGURE 30   Live-line tool disconnect switch. (Courtesy of Charles R. Miller)

FIGURE 31   Operating a disconnect switch. (© Matthew Collingwood/Shutterstock)

🔲 SHOCK PROTECTION

- This installation enables workers to operate the handle of the equipment quickly to de-energize the electrical service. However, some industrial facilities employ pad-mounted transformers with disconnecting devices that are operated using live-line tools (see Figure 32).

- Industrial workers must follow the same work processes and practices as utility workers.

FIGURE 32   Pad-mounted transformer. (Courtesy of Charles R. Miller)

🔌 SHOCK PROTECTION

Accessories are available that can equip the live-line tool to perform most functions. For example, the live-line tool can be utilized as a tool for rescue workers who might be in contact with an energized conductor, or the tool can be equipped with grounding equipment to discharge static voltage **(see Figure 33)**.

FIGURE 33   Static discharge stick. (Courtesy of Salisbury Electrical Safety, LLC)

🔌 ARC-FLASH PROTECTION

Arc-Flash Boundary (Arc-Flash and Arc-Blast Protection)

- The arc-flash boundary shall be the distance at which the incident energy equals 1.2 cal/cm² (5 J/cm²).

- Persons inside the arc-flash boundary must wear appropriate PPE, including arc-rated clothing.

Determining Available Fault Current

The level of fault current available in an electrical distribution system often can be furnished by the electric utility supplying the building or structure, or by the facility engineer/manager. Levels of available fault current also can be determined as follows:

- Perform the calculations by hand, as shown in Annex D of *NFPA 70E*.

- Use arc-flash calculation software.

- Hire an engineering/safety consultant.

Turning off the power is always the safest way to do electrical construction and maintenance work. Workers still need the correct PPE to test for the absence of voltage; until an electrically safe work condition has been established, a shock hazard and an arc-flash hazard exists. All electrical circuit conductors and circuit parts shall be considered energized until an electrically safe work condition has been established and verified. Therefore, safe work practices appropriate for the circuit voltage and energy level shall be used. Once an electrically safe work condition has been established, it is not necessary to determine the available fault current, because an arc-flash boundary does not exist.

ARC-FLASH PROTECTION

Determine Arc-Flash PPE

There are two methods permitted for the selection of PPE. Although either method is permitted, using both methods on the same piece of equipment is not permitted.

1. Incident Energy Analysis Method

An incident energy analysis is a component of an arc-flash risk assessment used to predict the incident energy of an arc flash for a specified set of conditions. An arc-flash risk assessment is a type of risk assessment. For information on estimating the incident energy, see Informative Annex D in *NFPA 70E*. For information on selection of arc-rated clothing and other PPE when the incident energy analysis method is used, see Table 130.5(G).

NOTE: Do not use the results of the incident energy analysis to specify an arc-flash PPE Category in Table 130.7(C)(15)(c).

- **See Table 3** in this *Ugly's* guide for selection of arc-rated clothing and other PPE when the incident energy analysis method is used.

2. Arc-Flash PPE Categories Method. The requirements of 130.7(C)(15) shall apply when the arc-flash PPE category method is used for the selection of arc-flash PPE.

When the arc-flash risk assessment performed in accordance with 130.5 indicates that arc-flash PPE is required and the arc-flash PPE category method is used for the selection of PPE for AC systems in lieu of the incident energy analysis of 130.5(G), Table 130.7(C)(15)(a) shall be used to determine the arc-flash PPE category. There are times when the arc-flash PPE categories method is not permitted. The incident energy analysis method shall be performed for any of the following:

- Power systems with greater than the estimated maximum available fault current

ARC-FLASH PROTECTION

TABLE 3   130.5(G) Selection of Arc-Rated Clothing and Other PPE When the Incident Energy Analysis Method Is Used

Incident energy exposures equal to 1.2 cal/cm² up to and including 12 cal/cm²

Arc-rated clothing with an arc rating equal to or greater than the estimated incident energy[a]

Arc-rated long-sleeve shirt and pants or arc-rated coverall or arc flash suit (SR)

Arc-rated face shield and arc-rated balaclava or arc flash suit hood (SR)[b]

Arc-rated outerwear (e.g., jacket, parka, rainwear, hard hat liner, high-visibility apparel) (AN)[e]

Heavy-duty leather gloves, arc-rated gloves, or rubber insulating gloves with leather protectors (SR)[c]

Hard hat

Safety glasses or safety goggles (SR)

Hearing protection

Leather footwear[d]

(continues)

TABLE 3 130.5(G) Selection of Arc-Rated Clothing and Other PPE When the Incident Energy Analysis Method Is Used—continued

Incident energy exposures greater than 12 cal/cm²

Arc-rated clothing with an arc rating equal to or greater than the estimated incident energy[a]

Arc-rated long-sleeve shirt and pants or arc-rated coverall or arc flash suit (SR)

Arc-rated arc flash suit hood

Arc-rated outerwear (e.g., jacket, parka, rainwear, hard hat liner, high-visibility apparel) (AN)[e]

Arc-rated gloves or rubber insulating gloves with leather protectors (SR)[c]

Hard hat

Safety glasses or safety goggles (SR)

Hearing protection

Leather footwear[d]

SR: Selection of one in group is required.

AN: As needed.

[a] Arc ratings can be for a single layer, such as an arc-rated shirt and pants or a coverall, or for an arc flash suit or a multilayer system if tested as a combination consisting of an arc-rated shirt and pants, coverall, and arc flash suit.

[b] Face shields with a wrap-around guarding to protect the face, chin, forehead, ears, and neck area are required by 130.7(C)(10)(c). Where the back of the head is inside the arc flash boundary, a balaclava or an arc flash hood shall be required for full head and neck protection.

[c] Rubber insulating gloves with leather protectors provide arc flash protection in addition to shock protection. Higher class rubber insulating gloves with leather protectors, due to their increased material thickness, provide increased arc flash protection.

[d] Footwear other than leather or dielectric shall be permitted to be used provided it has been tested to demonstrate no ignition, melting, or dripping at the estimated incident energy exposure.

[e] The arc rating of outer layers worn over arc-rated clothing as protection from the elements or for other safety purposes, and that are not used as part of a layered system, shall not be required to be equal to or greater than the estimated incident energy exposure.

⚡ ARC-FLASH PROTECTION

- Power systems with longer than the maximum fault clearing times

- Tasks with less than the minimum working distance

Determine Whether Arc-Flash PPE Is Required

- **See Table 4** in this *Ugly's* guide to identify when arc-flash PPE is required.

TABLE 4   130.5(C) Estimate of the Likelihood of Occurrence of an Arc-Flash Incident for AC and DC Systems

Task	Equipment Condition[a]	Likelihood of Occurrence[b]
Reading a panel meter while operating a meter switch. Performing infrared thermography and other noncontact inspections outside the restricted approach boundary. This activity does not include opening of doors or covers. Working on control circuits with exposed energized electrical conductors and circuit parts, nominal 125 volts AC or DC, or below without any other exposed energized equipment over nominal 125 volts AC or DC, including opening of hinged covers to gain access. Examination of insulated cable with no manipulation of cable. For DC systems, maintenance on a single cell of a battery system or multi-cell units in an open rack.	Any	No

(continues)

70

ARC-FLASH PROTECTION

TABLE 4 130.5(C) Estimate of the Likelihood of Occurrence of an Arc-Flash Incident for AC and DC Systems—continued

Task	Equipment Condition[a]	Likelihood of Occurrence[b]
For AC systems, work on energized electrical conductors and circuit parts, including voltage testing. Operation of a CB or switch the first time after installation or completion of maintenance in the equipment. For DC systems, working on energized electrical conductors and circuit parts of series connected battery cells, including voltage testing. Removal or installation of CBs or switches. Opening hinged door(s) or cover(s) or removal of bolted covers (to expose bare, energized electrical conductors and circuit parts). For DC systems, this includes bolted covers, such as battery terminal covers. Application of temporary protective grounding equipment, after voltage test. Working on control circuits with exposed energized electrical conductors and circuit parts, greater than 120 volts. Insertion or removal of individual starter buckets from motor control center (MCC).	Any	Yes

(continues)

71

TABLE 4   130.5(C) Estimate of the Likelihood of Occurrence of an Arc-Flash Incident for AC and DC Systems—continued

Task	Equipment Condition[a]	Likelihood of Occurrence[b]
Insertion or removal (racking) of circuit breakers (CBs) or starters from cubicles, doors open or closed.	Any	Yes
Insertion or removal of plug-in devices into or from busways.		
Examination of insulated cable with manipulation of cable.		
Working on exposed energized electrical conductors and circuit parts of equipment directly supplied by a panelboard or motor control center.		
Insertion or removal of revenue meters (kW-hour, at primary voltage and current).		
Insertion or removal of covers for battery intercell connector(s).		
For DC systems, working on exposed energized electrical conductors and circuit parts of utilization equipment directly supplied by a DC source.		
Opening voltage transformer or control power transformer compartments.		
Operation of outdoor disconnect switch (hookstick operated) at 1 kV through 15 kV.		

(continues)

ARC-FLASH PROTECTION

**TABLE 4   130.5(C) Estimate of the Likelihood of Occurrence
of an Arc-Flash Incident for AC and DC Systems—continued**

Task	Equipment Condition[a]	Likelihood of Occurrence[b]
Operation of outdoor disconnect switch (gang-operated, from grade) at 1 kV through 15 kV.	Any	Yes
Operation of a CB, switch, contactor, or starter. Voltage testing on individual battery cells or individual multicell units. Removal or installation of covers for equipment such as wireways, junction boxes, and cable trays that does not expose bare, energized electrical conductors and circuit parts. Opening a panelboard hinged door or cover to access dead front overcurrent devices. Removal of battery nonconductive intercell connector covers.	Normal	No
Maintenance and testing on individual battery cells or individual multicell units in an open rack. Insertion or removal of individual cells or multicell units of a battery system in an open rack.	Abnormal	Yes

(continues)

ARC-FLASH PROTECTION

TABLE 4   130.5(C) Estimate of the Likelihood of Occurrence of an Arc-Flash Incident for AC and DC Systems—continued

Task	Equipment Condition[a]	Likelihood of Occurrence[b]
Arc-resistant equipment with the DOORS CLOSED and SECURED, and where the available fault current and fault clearing time does not exceed that of the arc-resistant rating of the equipment in one of the following conditions: (1) Insertion or removal of individual starter buckets (2) Insertion or removal (racking) of CBs from cubicles (3) Insertion or removal (racking) of ground and test device (4) Insertion or removal (racking) of voltage transformers on or off the bus.	Abnormal	Yes

[a]Equipment is considered to be in a "normal operating condition" if all of the conditions in 110.4(D) are satisfied.

[b]As defined in this standard, the two components of risk are the likelihood of occurrence of injury or damage to health and the severity of injury or damage to health that results from a hazard. Risk assessment is an overall process that involves estimating both the likelihood of occurrence and severity to determine if additional protective measures are required. The estimate of the likelihood of occurrence contained in this table does not cover every possible condition or situation, nor does it address severity of injury or damage to health. Where this table identifies "No" as an estimate of likelihood of occurrence, it means that an arc flash incident is not likely to occur. Where this table identifies "Yes" as an estimate of likelihood of occurrence, it means an arc flash incident should be considered likely to occur. The likelihood of occurrence must be combined with the potential severity of the arcing incident to determine if additional protective measures are required to be selected and implemented according to the hierarchy of risk control identified in 110.5(H)(3).

ARC-FLASH PROTECTION

TABLE 4   130.5(C) Estimate of the Likelihood of Occurrence of an Arc-Flash Incident for *AC and DC Systems*—*continued*

Informational Note No. 1: An example of a standard that provides information for arc-resistant equipment referred to in Table 130.5(C) is IEEE C37.20.7, *Guide for Testing Switchgear Rated Up to 52 kV for Internal Arcing Faults.*

Informational Note No. 2: Improper or inadequate maintenance can result in increased fault clearing time of the overcurrent protective device, thus increasing the incident energy. Where equipment is not properly installed or maintained, PPE selection based on incident energy analysis or the PPE category method might not provide adequate protection from arc flash hazards.

Informational Note No. 3: Both larger and smaller available fault currents could result in higher incident energy. If the available fault current increases without a decrease in the fault clearing time of the overcurrent protective device, the incident energy will increase. If the available fault current decreases, resulting in a longer fault clearing time for the overcurrent protective device, incident energy could also increase.

Informational Note No. 4: The occurrence of an arcing fault inside an enclosure produces a variety of physical phenomena very different from a bolted fault. For example, the arc energy resulting from an arc developed in the air will cause a sudden pressure increase and localized overheating. Equipment and design practices are available to minimize the energy levels and the number of procedures that could expose an employee to high levels of incident energy. Proven designs such as arc-resistant switchgear, remote racking (insertion or removal), remote opening and closing of switching devices, high-resistance grounding of low-voltage and 5000-volt (nominal) systems, current limitation, and specification of covered bus or covered conductors within equipment are available to reduce the risk associated with an arc flash incident. See Informative Annex O for safety-related design requirements.

Informational Note No. 5: For additional direction for performing maintenance on overcurrent protective devices, see Chapter 2, Safety-Related Maintenance Requirements.

Informational Note No. 6: See IEEE 1584, *Guide for Performing Arc Flash Hazard Calculations*, for more information regarding incident energy and the arc flash boundary for three-phase systems.

Courtesy of National Fire Protection Association.

⚡ ARC-FLASH PROTECTION

Determine the Arc-Flash PPE Category

When arc-flash PPE is required, select the arc-flash PPE category from **Table 5A** or **Table 5B** in this *Ugly's* guide.

- These tables are applicable only when not exceeding the parameters for maximum available fault current, maximum fault clearing times, and minimum working distances.

- Arc-flash boundaries are listed for each of the equipment types.

- The arc-flash PPE category is based on the working distance listed with the type of equipment.

- Arc-Flash Hazard PPE Categories were formerly Hazard/Risk Categories.

TABLE 5A 130.7(C)(15)(a) Arc-Flash PPE Categories for Alternating Current (AC) Systems

Equipment	Arc-Flash PPE Category	Arc-Flash Boundary
Panelboards or other equipment rated 240 V and below Parameters: Maximum of 25 kA available fault current: maximum of 0.03 sec (2 cycles) fault clearing time; working distance 455 mm (18 in.)	1	485 mm (19 in.)
Panelboards or other equipment rated >240 V and up to 600 V Parameters: Maximum of 25 kA available fault current; maximum of 0.03 sec (2 cycles) fault clearing time; working distance 455 mm (18 in.)	2	900 mm (3 ft)
600 V class motor control centers (MCCs) Parameters: Maximum of 65 kA available fault current: maximum of 0.03 sec (2 cycles) fault clearing time; working distance 455 mm (18 in.)	2	1.5 m (5 ft)

(continues)

TABLE 5A   130.7(C)(15)(a) Arc-Flash PPE Categories for Alternating Current (AC) Systems—continued

Equipment	Arc-Flash PPE Category	Arc-Flash Boundary
600 V class motor control centers (MCCs) Parameters: Maximum of 42 kA available fault current: maximum of 0.33 sec (20 cycles) fault clearing time; working distance 455 mm (18 in.)	4	4.3 m (14 ft)
600 V class switchgear (with power circuit breakers or fused switches) and 600 V class switchboards Parameters: Maximum of 35 kA available fault current: maximum of up to 0.5 sec (30 cycles) fault clearing time; working distance 455 mm (18 in.)	4	6 m (20 ft)
Other 600 V class (277 V through 600 V, nominal) equipment Parameters: Maximum of 65 kA available fault current: maximum of 0.03 sec (2 cycles) fault clearing time; working distance 455 mm (18 in.)	2	1.5 m (5 ft)
NEMA E2 (fused contactor) motor starters, 2.3 kV through 7.2 kV Parameters: Maximum of 35 kA available fault current: maximum of up to 0.24 sec (15 cycles) fault clearing time; working distance 910 mm (36 in.)	4	12 m (40 ft)
Metal-clad switchgear, 1 kV through 15 kV Parameters: Maximum of 35 kA available fault current: maximum of up to 0.24 sec (15 cycles) fault clearing time; working distance 910 mm (36 in.)	4	12 m (40 ft)

(continues)

ARC-FLASH PROTECTION

TABLE 5A 130.7(C)(15)(a) Arc-Flash PPE Categories for Alternating Current (AC) Systems—continued

Equipment	Arc-Flash PPE Category	Arc-Flash Boundary
Metal enclosed interrupter switchgear, fused or unfused type construction, 1 kV through 15 kV Parameters: Maximum of 35 kA available fault current; maximum of 0.24 sec (15 cycles) fault clearing time; minimum working distance 910 mm (36 in.)	4	12 m (40 ft)
Other equipment 1 kV through 15 kV Parameters: Maximum of 35 kA available fault current; maximum of up to 0.24 sec (15 cycles) fault clearing time; minimum working distance 910 mm (36 in.)	4	12 m (40 ft)
Arc-resistant equipment up to 600-volt class Parameters: DOORS CLOSED and SECURED; with an available fault current and a fault clearing time that does not exceed the arc-resistant rating of the equipment*	N/A	N/A
Arc-resistant equipment 1 kV through 15 kV Parameters: DOORS CLOSED and SECURED; with an available fault current and a fault clearing time that does not exceed the arc-resistant rating of the equipment*	N/A	N/A

N/A: Not applicable
Note:
For equipment rated 600 volts and below and protected by upstream current-limiting fuses or current-limiting molded case circuit breakers sized at 200 amperes or less, the arc flash PPE category can be reduced by one number but not below arc flash PPE category 1.
*For DOORS OPEN, refer to the corresponding non–arc-resistant equipment section of this table.
Courtesy of National Fire Protection Association.

ARC-FLASH PROTECTION

TABLE 5B 130.7(C)(15)(b) Arc-Flash PPE Categories for Direct Current (DC) Systems

Equipment	Arc-Flash PPE Category	Arc-Flash Boundary
Storage batteries, DC switchboards, and other DC supply sources Parameters: Greater than or equal to 100 V and less than or equal to 250 V than or equal to 250 V Maximum arc duration and minimum working distance: 2 sec @ 455 mm (18 in.)		
Available fault current less than 4 kA	2	900 mm (3 ft)
Available fault current greater than or equal to 4 kA and less than 7 kA	2	1.2 m (4 ft)
Available fault current greater than or equal to 7 kA and less than 15 kA	3	1.8 m (6 ft)
Storage batteries, DC switchboards, and other DC supply sources Parameters: Greater than 250 V and less than or equal to 600 V Maximum arc duration and minimum working distance: 2 sec @ 455 mm (18 in.)		
Available fault current less than 1.5 kA	2	900 mm (3 ft)
Available fault current greater than or equal to 1.5 kA and less than 3 kA	2	1.2 m (4 ft)

(continues)

ARC-FLASH PROTECTION

TABLE 5B   130.7(C)(15)(b) Arc-Flash PPE Categories for Direct Current (DC) Systems—continued

Equipment	Arc-Flash PPE Category	Arc-Flash Boundary
Available fault current greater than or equal to 3 kA and less than 7 kA	3	1.8 m (6 ft)
Available fault current greater than or equal to 7 kA and less than 10 kA	4	2.5 m (8 ft)

Notes:

1. Apparel that can be expected to be exposed to electrolyte must meet both of the following conditions:

(a) Be evaluated for electrolyte protection

Informational Note: ASTM F1296, *Standard Guide for Evaluating Chemical Protective Clothing*, contains information on evaluating apparel for protection from electrolyte.

(b) Be arc-rated

Informational Note: ASTM F1891, *Standard Specification for Arc and Flame Resistant Rainwear*, contains information on evaluating arc-rated apparel.

2. A 2-second arc duration is assumed if there is no overcurrent protective device (OCPD) or if the fault clearing time is not known. If the fault clearing time is known and is less than 2 seconds, an incident energy analysis could provide a more representative result.

Courtesy of National Fire Protection Association.

Select the Appropriate PPE, Including Arc-Rated Clothing

• Once the appropriate arc-flash PPE category has been determined from *NFPA 70E* Table 130.7(C)(15)(a) or 130.7(C)(15)(b), use *NFPA 70E* Table 130.7(C)(15)(c) to select the appropriate personal protective equipment (PPE).

• **See Table 6** in this *Ugly's* guide to select the appropriate personal protective equipment (PPE).

NOTE: Do not use the results of the incident energy analysis to specify an arc-flash PPE Category in Table 130.7(C)(15)(c).

ARC-FLASH PROTECTION

TABLE 6   130.7(C)(15)(c) Personal Protective Equipment (PPE)

Arc-Flash PPE Category	PPE
1	**Arc-Rated Clothing, Minimum Arc Rating of 4 cal/cm²[a]** Arc-rated long-sleeve shirt and pants or arc-rated coverall Arc-rated face shield[b] or arc-flash suit hood Arc-rated jacket, parka, high-visibility apparel, rainwear, or hard hat liner (AN)[f] **Protective Equipment** Hard hat Safety glasses or safety goggles (SR) Hearing protection (ear canal inserts)[c] Heavy duty leather gloves, arc-rated gloves, or rubber insulating gloves with leather protectors (SR)[d] Leather footwear[e] (AN)
2	**Arc-Rated Clothing, Minimum Arc Rating of 8 cal/cm²[a]** Arc-rated long-sleeve shirt and pants or arc-rated coverall Arc-rated flash suit hood or arc-rated face shield[b] and arc-rated balaclava Arc-rated jacket, parka, high-visibility apparel, rainwear, or hard hat liner (AN)[f] **Protective Equipment** Hard hat Safety glasses or safety goggles (SR) Hearing protection (ear canal inserts)[c] Heavy-duty leather gloves, arc-rated gloves, or rubber insulating gloves with leather protectors (SR)[d] Leather footwear[e]

(continues)

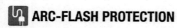

**TABLE 6   130.7(C)(15)(c) Personal Protective
Equipment (PPE) —continued**

Arc-Flash PPE Category	PPE
3	**Arc-Rated Clothing Selected so That the System Arc Rating Meets the Required Minimum Arc Rating of 25 cal/cm[2a]** Arc-rated long-sleeve shirt (AR) Arc-rated pants (AR) Arc-rated coverall (AR) Arc-rated arc-flash suit jacket (AR) Arc-rated arc-flash suit pants (AR) Arc-rated arc-flash suit hood Arc-rated gloves or rubber insulating gloves with leather protectors (SR)[d] Arc-rated jacket, parka, high-visibility apparel, rainwear, or hard hat liner (AN)[f] **Protective Equipment** Hard hat Safety glasses or safety goggles (SR) Hearing protection (ear canal inserts)[c] Leather footwear[e]

(continues)

TABLE 6 130.7(C)(I5)(c) Personal Protective Equipment (PPE)—continued

Arc-Flash PPE Category	PPE
4	**Arc-Rated Clothing Selected So That the System Arc Rating Meets the Required Minimum Arc Rating of 40 cal/cm²ᵃ**
	Arc-rated long-sleeve shirt (AR)
	Arc-rated pants (AR)
	Arc-rated coverall (AR)
	Arc-rated arc-flash suit jacket (AR)
	Arc-rated arc-flash suit pants (AR)
	Arc-rated arc-flash suit hood
	Arc-rated gloves or rubber insulating gloves with leather protectors (SR)ᵈ
	Arc-rated jacket, parka, high-visibility apparel, rainwear, or hard hat liner (AN)ᶠ
	Protective Equipment
	Hard hat
	Safety glasses or safety goggles (SR)ᶜ
	Hearing protection (ear canal inserts)ᶜ
	Leather footwearᵉ

AN: as needed (optional). AR: as required. SR: selection required.

ᵃArc *rating is* defined in Article 100 of *NFPA 70E.*

ᵇFace shields are to have wrap-around guarding to protect not only the face but also the forehead, ears, and neck, or alternatively, an arc-rated arc-flash suit hood is required to be worn.

ᶜOther types of hearing protection are permitted to be used in lieu of or in addition to ear canal inserts provided they are worn under an arc-rated arc-flash suit hood.

ᵈRubber insulating gloves with leather protectors provide arc flash protection in addition to shock protection. Higher class rubber insulating gloves with leather protectors, due to their increased material thickness, provide increased arc flash protection.

ᵉFootwear other than leather or dielectric shall be permitted to be used provided it has been tested to demonstrate no ignition, melting or dripping at the minimum arc rating for the respective arc flash PPE category.

ARC-FLASH PROTECTION

REMEMBER: Arc-flash PPE categories apply until an electrically safe work condition has been established and verified. After the eight steps in 70E 120.5 have been verified and an electrically safe work condition has been established, there is no arc-flash hazard category.

Arc-Rated Clothing

Eliminating the risk of clothing ignition or melting prevents many injuries and reduces the severity of the remainder. Arc-rated clothing can reduce the severity of burn injuries. At the very least, wearing arc-rated clothing would eliminate the risk associated with clothing ignition

Clothing that meets the requirements of the current edition of *ASTM F1506, Standard Performance Specification for Flame Resistant and Arc Rated Protective Clothing Worn by Workers Exposed to Flames and Electric Arc* is considered arc-rated clothing. Fabric or other material that is flame resistant does not ignite. The flame-resistant characteristic remains with the clothing for the life of the garment. For a list of standards that regulate personal protective equipment (PPE), **see Table 7**.

 **ARC-FLASH PROTECTION**

TABLE 7 130.7(C)(14) Informational Note: Standards for Personal Protective Equipment

Subject	Document Title	Document Number
Clothing—Arc Rated	Standard Performance Specification for Flame Resistant and Arc Rated Textile Materials for Wearing Apparel for Use by Electrical Workers Exposed to Momentary Electric Arc and Related Thermal Hazards	ASTM F1506
	Standard Guide for Industrial Laundering of Flame, Thermal, and Arc Resistant Clothing	ASTM F1449
	Standard Guide for Home Laundering Care and Maintenance of Flame, Thermal and Arc Resistant Clothing	ASTM F2757
Aprons—Insulating	Live working—Protective clothing against the thermal hazards of an electric arc—Part 1-1: Test methods—Method 1: Determination of the arc rating (ELIM, ATPV, and/or EBT) of clothing materials and of protective clothing using an open arc	IEC 61482–1–1
	Live working—Protective clothing against the thermal hazards of an electric arc—Part 2: Requirements	IEC 61482–2
	Standard Specification for Electrically Insulating Aprons	ASTM F2677

(continues)

ARC-FLASH PROTECTION

TABLE 7 Table 130.7(C)(14) Informational Note: Standards for Personal Protective Equipment—continued

Subject	Document Title	Document Number
Eye and Face Protection—General	American National Standard for Occupational and Educational Professional Eye and Face Protection	ANSI/ ISEA Z87.1
Face—Arc Rated	Standard Test Method for Determining the Arc Rating and Standard Specification for Eye or Face Protective Products	ASTM F2178
Fall Protection	Standard Specification for Personal Climbing Equipment	ASTM F887
Footwear—Dielectric Specification	Standard Specification for Dielectric Footwear	ASTM F1117
Footwear—Dielectric Test Method	Standard Test Method for Determining Dielectric Strength of Dielectric Footwear	ASTM F1116
Footwear—Standard Performance Specification	Standard Specification for Performance Requirements for Protective (Safety) Toe Cap Footwear	ASTM F2413
Footwear—Standard Test Method	Standard Test Methods for Foot Protections	ASTM F2412
Gloves—Arc Rated	Standard Test Method for Determining Arc Ratings of Hand Protective Products Developed and Used for Electrical Arc Flash Protection	ASTM F2675/ F2675M
Gloves—Leather Protectors	Standard Specification for Leather Protectors for Rubber Insulating Gloves and Mittens	ASTM F696

(continues)

TABLE 7   Table 130.7(C)(14) Informational Note: Standards for Personal Protective Equipment—continued

Subject	Document Title	Document Number
Gloves—Rubber Insulating	Standard Specification for Rubber Insulating Gloves	ASTM D120
Gloves and Sleeves—In-Service Care	Standard Specification for In-Service Care of Insulating Gloves and Sleeves	ASTM F496
Head Protection—Hard Hats	American National Standard for Head Protection	ANSI/ISEA Z89.1
Rainwear—Arc Rated	Standard Specification for Arc and Flame Resistant Rainwear	ASTM F1891
Rubber Protective Products—Visual Inspection	Standard Guide for Visual Inspection of Electrical Protective Rubber Products	ASTM F1236
Sleeves—Insulating	Standard Specification for Rubber Insulating Sleeves	ASTM D1051

NOTE: ASTM—American Society for Testing and Materials. ANSI—American National Standards Institute. ISEA—International Safety Equipment Association

Courtesy of National Fire Protection Association.

"Arc rating" is defined as the value attributed to materials that describes their performance to exposure to an electrical arc discharge. The arc rating is expressed in cal/cm^2 and is derived from the determined value of the arc thermal performance value (ATPV) or energy of breakopen threshold (E_{BT}) (should a material system exhibit a breakopen response below the ATPV value).

- Protective clothing and equipment must have a rating equal to or higher than the prospective incident energy exposure.

- Arc rating is reported as either ATPV or E_{BT}, whichever is the lower value.

- Arc-rated clothing or equipment indicates that it has been tested for exposure to an electric arc.

⚡ ARC-FLASH PROTECTION

- Flame-resistant (FR) clothing without an arc rating has *not* been tested for exposure to an electric arc.

- All arc-rated clothing is also flame resistant.

ATPV is defined in *ASTM F1959/F1959M, Standard Test Method for Determining the Arc Rating of Materials for Clothing,* as the incident energy (cal/cm²) on a material or a multilayer system of materials that results in a 50% probability that sufficient heat transfer through the tested specimen is predicted to cause the onset of a second-degree skin burn injury, based on the Stoll curve.

- An alternative to ATPV is energy of breakopen threshold (E_{BT}).

- If the ATPV cannot be reached (or the material exhibits a breakopen response below the ATPV value), the energy required to break open the test fabric (E_{BT}) is used as the value for the arc rating.

E_{BT} is defined in *ASTM F1959/F1959M, Standard Test Method for Determining the Arc Rating of Materials for Clothing,* as the incident energy (cal/cm²) on a material or a material system that results in a 50% probability of breakopen. Breakopen is defined as a hole with an area of 1.6 cm² (0.5 in²) or an opening of 2.5 cm (1.0 in.) in any dimension.

- It is permissible to use the ATPV or E_{BT} value to select arc-rated clothing.

- Breakopen is a material response evidenced by the formation of one or more holes in the innermost layer of arc-rated material that would allow flame to pass through the material.

Label Requirements

ASTM F1506 requires the clothing manufacturer to provide certain information on the label in each garment **(see Figure 34)**. The information is intended to provide a worker with sufficient information to properly select and use arc-rated protection. The label must contain at least the following information:

⚡ ARC-FLASH PROTECTION

FIGURE 34   Clothing label. (Courtesy of Oberon Company)

- A tracking code

- Indication that the garment conforms to *ASTM F1506*

- Manufacturer's name

- Size and associated standard label information

- Care instructions and fiber content

- Arc rating, either ATPV or E_{BT}

- Care instructions

Make sure each item of arc-rated clothing is suitable for the expected exposure by comparing the arc rating on the clothing label with the incident energy (cal/cm²) exposure level.

TCG40™ Electrical Arc Flash Hood

For electric arc exposures, perform an electric arc hazard assessment and wear the properly rated flame resistant clothing

Use this hood with proper respiratory protection (if required)
Use this **Hood** with suitably arc rated **Overall** or **Coat & Pants/Bib-Overall**

Meets ASTM F1506²⁰ᵃ, NFPA 70E¹⁵ & CSA Z462¹⁵

Arc Rating **42** cal/cm² (ATPV)

Hood tested in accordance with ASTM F2178-12
Fabric System tested in accordance with ASTM F1959-14

100% Aramid

Caution
Flammable contaminants will reduce the thermal performance of any flame resistant garment. Wash garment to ensure that no greases, oily soils or other flammable contaminants are present when garment is worn. Repairs to the garment must be made with the same thread and fabric.

Do not reuse this product after an arc exposure

Washing Instructions

Window: Remove window from Hood shell. Rinse window with warm water. Tap dry inner surface w/soft cloth. Allow inner antifog surface to regenerate.
DO NOT USE cleaners or disinfectants on inner surface!

Fabric: WASH – Wash this garment separately from other types of garments or fabrics. Use warm water & mild detergent. DO not use detergents or additives containing (or creating) chlorine bleach or oxygen bleach (example: Hydrogen Peroxide) or enzymes. Do not use soap or fabric softeners. Do not wring dry.

DRY – Tumble dry low heat and remove immediately. Do not iron. Garment can be dry cleaned. Do not use disinfectants

Important Warnings
This product does not provide insulation from electric shock. This product, when properly selected and worn, is designed to provide protection from burn injuries resulting from an electrical arc flash in accordance with ASTM F1506 & NFPA 70E. It is important to note that this product will provide limited or no protection against sound, pressure, projectiles and respiratory hazards which may result from an arc incident.
Face protective products which are effective against the arc flash hazard can reduce visibility. NFPA 70E calls for the use of additional lighting if work conditions require.

Remember
Always perform a hazard analysis to determine the potential arc hazard energy level. Only use PPE with an arc rating (ATPV or EBT) that is equal to or greater than your potential arc hazard level.

⬣OBERON
OBERON COMPANY
Made in U.S.A
visit us on the web @ www.arcflash.com
TCG40F-HOOD

R17-40

‖‖‖‖‖‖‖‖‖‖‖‖‖‖‖
475745722703

🔌 ARC-FLASH PROTECTION

- When a worker performs the arc-flash risk assessment required by *NFPA 70E*, they will know the incident energy (cal/cm^2) level of potential exposure; the working distance will also be known.

- By reading the label in the arc-rated clothing, the worker knows what level of thermal protection can be expected from the arc-rated clothing.

- If the arc rating of the equipment listed on the clothing label meets or exceeds the expected exposure at the working distance, the arc-rated clothing is acceptable for that work task.

Thermal Barrier

- Electrical shock protection in the form of rated rubber products reduces the risk of electrical shock or electrocution by reducing the amount of current that might flow in a worker's body.

- The rubber products introduce an additional barrier that resists the flow of electrical current.

- Likewise, arc-rated clothing reduces the risk of thermal burn by reducing the amount of thermal energy that might flow onto a worker's body.

- The arc-rated clothing introduces an additional barrier, which resists the flow of thermal energy.

Selection and Use

The most severe injuries from exposure to an electrical arc are the result of flammable or meltable clothing igniting or melting. In general, the duration of an electrical arc is limited by the clearing time of the overcurrent device. Consensus standards and codes guide the selection of overcurrent devices.

- Overcurrent devices function by monitoring the amount of current flowing in the circuit; when the amount of current exceeds the trip setting of the overcurrent device, the circuit is opened, and the source of energy is removed.

- Contemporary codes and standards permit overcurrent devices to be large to reduce the chance of nuisance trips.

- Architects and engineers generally specify circuit breakers and fuses that will reduce the risk of fire in the building and protect the equipment from destruction should a fault occur. Fire was one of the first hazards associated with the use of electrical energy and remains a primary concern.

Electrical equipment usually is constructed from metal components such as copper or aluminum conductors, electrical insulating components, and steel structures or enclosures.

- The metal components of electrical equipment remain solid until the temperature is elevated to the melting point.

- A slightly higher temperature results in the liquid metal beginning to evaporate by boiling, becoming metal vapor.

- The change of state of the metal components begins when the temperature is in the range of 982°C (1,800°F). Within a few hundred degrees, the metal components evaporate and become metal vapor.

- Copper expands by a factor of 67,000 times when it turns from a solid to a vapor.

- Human tissue begins to be destroyed when the temperature is elevated to about 66°C (150°F) and held at that temperature for 1 second.

- Cells are destroyed in 1/10 of 1 second when the temperature of the tissue is elevated to about 93°C (200°F).

- To avoid a burn injury, then, both the temperature and the duration of the exposure must be limited **(see Table 8)**.

ARC-FLASH PROTECTION

TABLE 8   Effect of Temperature on Human Skin

Skin Temperature	Time to Reach Temperature	Damage Caused
43°C (110°F)	6.0 hours	Cell breakdown begins
70°C (158°F)	1.0 second	Total cell destruction
80°C (176°F)	0.1 second	Second-degree burn
93°C (200°F)	0.1 second	Third-degree burn

Ray A., & Jane G. Jones. (2000). *Electrical Safety in the Workplace*, Sudbury, MA: Jones and Bartlett Publishers.

Limiting Fault-Current Time

- Overcurrent devices remove the source of energy from the arc in a predetermined period of time. Overcurrent devices do not clear the arcing fault instantaneously.

- When a fuse element begins to melt, current continues to flow until the opening in the melting element is large enough to break the circuit.

- Fuses contain a substance similar to sand, which flows into the opening in the melting element and quenches the arc.

- If the fuse is a current-limiting fuse and the fault current is large, the element begins to melt in one-quarter of one cycle, or about 4.5 milliseconds, and clears the fault in less than two cycles, or 34 milliseconds.

- If the current in the arcing fault is below the current-limiting range of the fuse; the arcing fault might continue for several seconds.

⏻ ARC-FLASH PROTECTION

Protective Clothing Requirements

To avoid injury, select protective clothing that has the following characteristics:

- The clothing must not ignite or continue to burn after the arcing fault has been removed.

- The clothing must not melt onto or into the worker's skin as a result of being exposed to the arc.

- The clothing must provide sufficient thermal insulation to prevent the worker's skin tissue from being heated to destruction.

The Arc-Flash Boundary

One primary purpose of an arc-flash risk assessment is to determine the arc-flash boundary.

- The label on the equipment should identify the arc-flash boundary, either in meters or feet and inches.

- Determine the arc-flash boundary from the label on the equipment. The arc-flash boundary of an idealized arc in free air is a spherical shape, measured from the potential arc to any part of a worker's body **(see Figure 35)**.

- When any part of a worker's body is close to the source of a potential arc, a thermal burn injury is possible, and the worker's body must be protected from a possible burn.

- When arc-rated clothing is necessary for a work task, all buttons, zippers, and other closing mechanisms must be completely closed.

- The top fastener near the worker's neck must be closed to minimize the chance that high-temperature gases could get behind the clothing and breach the protective characteristic.

ARC-FLASH PROTECTION

FIGURE 35 Arc-flash boundary. (Courtesy of Charles R. Miller)

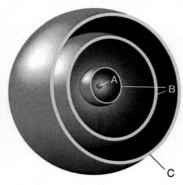

A. Exposed energized electrical conductor or circuit part

B. Shock protection boundaries

C. Arc-flash boundary

When an arc-flash hazard exists, an approach limit at a distance from a prospective arc source within which a person could receive a second-degree burn if an electrical arc flash were to occur is the arc-flash boundary. Note, the arc-flash boundary is not always the boundary farthest away from the energized electrical conductor or circuit part.

- *ASTM F1506* ensures that adequate layers of arc-rated material protect fasteners on all arc-rated clothing from the extreme heat that could cause a worker to be burned.

- The outer layer of the worker's protective clothing must be arc rated.

- The arc rating of any label or identifying patch must equal or exceed the arc rating of the garment.

- Nonmelting, flammable fiber garments shall be permitted to be used as underlayers in conjunction with arc-rated garments in a layered system.

- However, if nonmelting, flammable fiber garments are used as underlayers, the system arc rating shall be sufficient to prevent

ARC-FLASH PROTECTION

breakopen of the innermost arc-rated layer at the expected arc exposure incident energy level to prevent ignition of flammable underlayers.

- The head and face of a worker who must perform a task within the arc-flash boundary are exposed to the potential arc flash.

- If the worker's head is within the arc-flash boundary, they should select and wear protective equipment that will eliminate or minimize the risk of exposure to air or gases that are very hot. Protective equipment that covers the head entirely is available.

- Arc-rated face shields and balaclavas are also available. Note, however, that an arc-rated balaclava also requires face protection **(see Figure 36)**.

- An arc-rated balaclava shall be used with an arc-rated face shield when the back of the head is within the arc-flash boundary. An arc-rated hood shall be permitted to be used instead of an arc-rated face shield and balaclava.

FIGURE 36   Arc-rated balaclava from Oberon with an ATPV rating of 14 cal/cm². (Courtesy of Billy Huff)

⚡ ARC-FLASH PROTECTION

- An arc-rated hood shall be used when the anticipated incident energy exposure exceeds 12 cal/cm².

- Equipment intended to protect a worker's head and face is assigned an arc rating that mirrors the arc rating assigned to arc-rated clothing.

- When performing the necessary arc-flash risk assessment, consider the location of the work task in relation to the different parts of the body.

- If the work task is near the floor, be aware that hot air and gases could get behind the protective equipment at the lower extremity.

- Injuries to the legs are possible from the entrance of hot air and gases under the leg protection on the clothing.

- Make sure that protection for the bottom lower extremities prevents hot air and gases from getting behind it.

- If the work task (and potential arcing fault) is at an elevated position, the risk of hot air and gases getting behind the protective equipment near the feet is reduced.

- Clothing shall cover potentially exposed areas as completely as possible. Shirt and coverall sleeves shall be fastened at the wrists, shirts shall be tucked into pants, and shirts, coveralls, and jackets shall be closed at the neck.

Testing Arc-Rated Fabrics

- Arc-rated fabrics are tested as defined by *ASTM F1506*.

- Representative samples of the fabrics are subjected to the vertical flame test described in *ASTM D6413/D6413M, Standard Test Method for Flame Resistance of Textiles (Vertical Test)*.

- This test determines that the fabric will not ignite and burn.

🔌 ARC-FLASH PROTECTION

- The material is permitted to continue to flame for a period not exceeding 2 seconds after the test flame is removed and 5 seconds after an arc test.

- The char length in the vertical flame test is not permitted to exceed 6 inches.

- The test is conducted on new fabric and repeated on fabric that has been laundered 25 times.

- After the flammability of the fabric has been established, the fabric is then subjected to an arc test as described in *ASTM F1959/ F1959M, Standard Test Method for Determining the Arc Rating of Materials for Clothing.*

- An arc rating is established for each fabric sample when the *F1959/ F1959M* test is conducted. The afterflame of the fabric must not exceed 5 seconds.

- Some fabrics exhibit an "afterglow" that exceeds 5 seconds. However, the afterglow is a property of the fabric and is not considered a flame.

- The product cannot exhibit any indication of melting or dripping when either the arc test or the vertical flame test is conducted.

- The arc rating determines the protective characteristics of the fabric.

- When the product is sold to protect workers from arcing faults, clothing manufacturers are required to provide the arc rating on the label.

- Clothing manufacturers are also permitted to indicate the arc rating on the clothing's surface, such as a shirt pocket or sleeve.

🔌 ARC RATING

ASTM F1506, Standard Performance Specification for Flame Resistant and Arc Rated Protective Clothing Worn by Workers Exposed to Flames and Electric Arcs is a performance specification that identifies specific tests and acceptable results.

- The specification covers fabrics that are woven or knitted. The specification is not intended for use with other fabric construction.

- In general, fabric is measured in weight per unit of area.

- Fabric produced for use as arc-rated clothing is also measured in ounces per square yard. As the weight per square yard of arc-rated fabric increases, the thermal insulating ability also increases.

- In most cases, fabric from different clothing manufacturers is constructed from different textile products and blends of products from different textile manufacturers.

- The weight of the arc-rated protective clothing is not an accurate indicator of the degree of protection provided by the clothing. The arc rating is the only reasonable indicator of the protection provided by the clothing.

- Employers have a common practice of providing patches with an employee's name or a company logo. Employees are then encouraged to attach their name and company logo to their shirt or uniform.

- Any patch or logo attached to arc-rated clothing also must be arc rated. Patches that are flammable or meltable provide fuel that can ignite or melt and increase the effect of an arcing fault.

- The protective characteristic of arc-rated clothing depends on the surface of the fabric being clean and free from flammable material such as grease and oil.

🔌 ARC RATING

- Wiping cloths should not be kept on a worker's body or clothing pocket when they are at risk of being exposed to a potential arcing fault. Just as a flammable patch adds fuel to the fire, grease and oil also add fuel to the fire.

- Wiping cloths that are kept in a pocket are likely to ignite and burn if exposed to an arcing fault. A burning wiping cloth in a pocket is likely to overcome the protection provided by the arc-rated clothing.

 INSULATING FACTOR OF LAYERS

Air is a relatively good thermal insulator.

- If two or more layers of arc-rated fabric are used as part of a protective system, a layer of air is trapped between the layers.

- The air layer provides additional thermal insulation between a worker's skin and any potential arcing fault. (See Table 6, *NFPA 70E* Table 130.7(C)(15)(c), for PPE Categories and Required Minimum Arc Rating.)

The basic idea of providing protection is adding thermal insulation between a worker and an arcing fault. As indicated above, a layer of air increases the thermal protection. Arc-rated clothing, then, should be loose fitting, but not so loose as to interfere with the worker's movement. When the arc-rated clothing is in contact with the worker's skin, thermal energy can be conducted through the fabric to the worker.

- Fabric used to construct arc-rated clothing provides only thermal insulation. It provides no protection from shock or electrocution. However, arc-rated protective equipment does not increase the chance of electrical shock unless the material is metalized or contains a conductive component, such as carbon. Although the reflectivity of a metalized garment would reduce the amount of energy conducted through the garment, the conductive garment would increase the chance of initiating an arcing fault. Some protective clothing contains a small amount of carbon to reduce the effect of static electricity. Small amounts of carbon embedded in the thread do not increase exposure to shock. Users should contact the clothing or fabric manufacturer for further details.

- An arc-rated arc-flash suit hood provides greater protection than a face shield. In addition to the thermal protection provided by an arc-rated hood, protection is also provided by a small quantity of air trapped in the hood. Should an arcing fault occur, the uncontaminated air inside the hood might be an advantage for a worker **(see Figure 37)**.

INSULATING FACTOR OF LAYERS

FIGURE 37 Arc-rated arc flash suit hood with an ATPV rating of 67 cal/cm². (Courtesy of Oberon Company)

- A face shield provides protection from energy that is transmitted in the form of radiation. Energy that is transmitted by convection could flow behind the face shield and cause a burn. A face shield provides protection from flying parts and pieces that may be expelled by the blast component of the arcing fault.

NOTE: Eye protection (safety glasses or goggles) shall always be worn under face shields or hoods.

PPE CONFIGURATIONS

PPE can consist of many different configurations, including various combinations of arc-rated clothing and protective equipment **(see Figures 38–40)**.

- A worker might choose to wear coveralls or a shirt-and-pants combination.

- A typical layering system might include cotton underwear, a cotton shirt and trouser, and an arc-rated coverall.

FIGURE 38   Arc-rated shirts and pants. (Courtesy of Oberon Company)

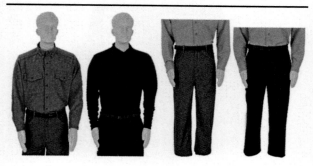

FIGURE 39   Arc-rated coveralls. (Courtesy of Oberon Company)

FIGURE 40   Complete arc-flash suit. (Courtesy of Oberon Company)

⚡ PPE CONFIGURATIONS

- Specific tasks might call for additional arc-rated layers to achieve the required protection level.

- The PPE might consist of a single layer, or it might consist of two or more layers.

- The PPE might consist of both natural fiber material and arc-rated material. The manufacturer should be consulted to determine the overall rating of any multiple-layer protective system.

Do not wear clothing made from fabrics that melt at temperatures below 315°C (600°F), such as polyester, nylon, acetate, acrylic, polyethylene, polypropylene, and spandex. Clothing containing blends of these products must not be worn unless the product is assigned an arc rating by the manufacturer.

Many undergarments are made of polyester, nylon, acetate, and similar products. If an undergarment melts, the melted fabric will destroy any skin tissue that it contacts. Workers who wear arc-rated protective equipment should wear undergarments of cotton or arc-rated fabric only.

NOTE: An incidental amount of elastic used on nonmelting fabric underwear or socks shall be permitted.

AN ARCING FAULT CONVERTS ELECTRICAL ENERGY INTO OTHER FORMS OF ENERGY

Arc-rated products discussed in this section provide protection from the energy that is converted into thermal energy. All electrical arcs also convert some energy into pressure. The pressure wave created when an arcing fault is initiated produces a wave that results in a force known as an arc blast.

In past editions of *NFPA 70E*, an informational note stated greater emphasis might have been necessary with respect to de-energizing where exposed to electrical hazards when incident energy exceeded 40 cal/cm² at the working distance. The informational note that was deleted in the 2018 edition seemed to be saying the blast pressure could be dangerous when incident energy exceeded 40 cal/cm². Evidence now shows there is no direct correlation between incident energy levels and blast pressure. Blast pressure is more of a function of short-circuit current.

The important thing to learn from 130.7(A) Informational Note No. 3 being deleted is that greater emphasis should **always** be placed on de-energizing regardless of the incident energy level.

CLOTHING DESIGNED ESPECIALLY FOR WOMEN

Until a few years ago, arc-rated clothing was made in sizes for male workers only, although women might wear it. Several manufacturers such as ArcWear™, Tyndale, and Workrite® now offer arc-rated clothing, including both outerwear and underwear, in standard sizes for women.

Men can wear cotton underwear under their arc-rated clothing with no danger of it melting. However, some women's underwear is made from polyester or similar materials that could melt in an arc-flash event, and some bras contain metal underwires, hooks, or clips that should not be worn (even under arc-rated clothing) if there is potential for exposure to the arc-flash hazard.

⚡ FOOT PROTECTION

Leather footwear or dielectric footwear or both provide some arc flash protection to the feet and shall be used in all exposures greater than 4 cal/cm² (16.75 J/cm²) **(see Figure 41)**.

Footwear other than leather or dielectric shall be permitted to be used provided it has been tested to demonstrate no ignition, melting or dripping at the estimated incident energy exposure or the minimum arc rating for the respective Arc Flash PPE Category.

FIGURE 41 Leather Footwear. (Courtesy of Charles R. Miller)

⚡ PROTECTIVE MATERIALS

- Leather products are intended to prevent abrasion and penetration.

- Although leather provides sound protection from abrasion and from penetration by parts that are static, moving parts and pieces ejected by an arc blast will only be partially impeded. Some arc-rated clothing contains material (such as Kevlar®) that is normally used to stop small flying objects such as bullets.

- If the arc-rated clothing successfully stops the moving part, the energy contained in the moving object will be transferred to the person. Face shields, safety glasses, safety goggles, and viewing windows in hoods are intended to provide protection from impact. However, these products are not tested in an environment where objects of significant size and momentum might exist.

- Rubber products worn for protection from electrical shock provide a layer that can resist penetration and abrasion. Again; however, energy contained in the flying parts and pieces is transferred to the person, even if the rubber protection is not penetrated.

The bottom line is that no product exists that can protect a person from the blast effects of an arcing fault. Products worn for protection from other electrical hazards can provide some resistance to penetration, but they do not absorb the kinetic energy in the momentum of flying parts and pieces. Only two options exist to protect from a potential arc blast:

1. The equipment must be placed in an electrically safe work condition.

2. There must be sufficient distance from the source of the potential arc.

🔌 PROTECTIVE MATERIALS

NFPA 70E suggests that a safe electrical work environment has three components:

- *NEC*-compliant installation

- Proper maintenance

- Safe work practices

Condition of maintenance is defined in *70E* as the state of the electrical equipment considering the manufacturers' instructions; manufacturers' recommendations; and applicable industry codes, standards, and recommended practices.

Adequate maintenance of electrical systems improves reliability and reduces employee exposure to electrical hazards. The following examples illustrate this axiom:

- When overcurrent devices (circuit breakers and fuses) function as designed to clear faults quickly, arc-flash and arc-blast hazards are reduced.

- When grounding and bonding systems function properly, shock hazard and the sensation of "touch voltage" are reduced for employees in the workplace.

Qualified Persons

Only qualified persons are permitted to perform maintenance on electrical equipment, circuits, and installations.

- A qualified person is defined as one who has demonstrated skills and knowledge related to the construction and operation of electrical equipment and installations and has received safety training to identify the hazards and reduce the associated risk.

# GENERAL ELECTRICAL EQUIPMENT

Chapter 2 of *NFPA 70E* contains general requirements for maintaining electrical equipment. It refers to *NFPA 70B, Recommended Practice for Electrical Equipment Maintenance*, for specific maintenance methods and tests. Chapter 2 is divided into 10 articles.

Article 205—General Maintenance Requirements

Article 210—Substations, Switchgear Assemblies, Switchboards, Panelboards, Motor Control Centers, and Disconnect Switches

Article 215—Premises Wiring

Article 220—Controller Equipment

Article 225—Fuses and Circuit Breakers

Article 230—Rotating Equipment

Article 235—Hazardous (Classified) Locations

Article 240—Batteries and Battery Rooms

Article 245—Portable Electric Tools and Equipment

Article 250—Personal Safety and Protective Equipment

Chapter 2 of *NFPA 70E* covers practical safety-related maintenance requirements for electrical equipment and installations in workplaces as included in *NFPA 70E* 90.2. Requirements in Chapter 2 identify only maintenance directly associated with employee safety. Chapter 2 does not prescribe specific maintenance methods or testing procedures. It is left to the employer to choose from the various maintenance methods available to satisfy the requirements of Chapter 2 in NFPA 70E.

For the purpose of Chapter 2, maintenance shall be defined as preserving or restoring the condition of electrical equipment and installations, or parts of either, for the safety of employees who work where exposed to electrical hazards. Repair or replacement of individual portions or parts of equipment shall be permitted without requiring modification or replacement of other portions or parts that are in a safe condition.

Note: Refer to NFPA 70B, *Recommended Practice for Electrical Equipment Maintenance*; ANSI/NETA MTS, *Standard for Maintenance Testing Specifications for Electrical Power Distribution Equipment and Systems*; and IEEE 3007.2, *Recommended Practice for the Maintenance of Industrial and Commercial Power Systems*, for guidance on maintenance frequency, methods, and tests.

 SPECIAL ELECTRICAL EQUIPMENT

NFPA 70E also includes safety-related work practices for installing and maintaining some specific types of electrical equipment. Electricians and electrical contractors doing typical residential-commercial-industrial jobs rarely work on these specialized types of equipment. *NFPA 70E*, Chapter 3, covers the following:

1. Electrolytic cells

2. Batteries and battery rooms

3. Lasers

4. Power electronic equipment

5. Research and development (R&D) laboratories

6. Capacitors

SPECIAL ELECTRICAL EQUIPMENT

1. Electrolytic Cells (Article 310)

An electrolytic cell is defined in *NFPA 70*, National Electrical Code as a tank or vat in which electrochemical reactions are caused by applying electric energy for the purpose of refining or producing usable materials. Electrolytic cells are industrial equipment used in the production of metals such as copper, aluminum, and zinc. Each cell (or pot) is a direct-current (DC) electric furnace, and multiple DC furnaces connected in series are called a *cell line*. The working area surrounding this equipment is called a *cell-line working zone.*

Heat Causes Special Hazards

Temperatures inside electrolytic cells can reach 900°C (1,652°F), and operating voltages can reach 1,000 volts or more. This combination can make conventional PPE (such as arc-flash suits) impractical. Using regular *NFPA 70E* rules to protect qualified persons against available arc-flash incident energy can incapacitate them through overheating.

Operation and maintenance of electrolytic cell lines might require contact by employees with exposed energized surfaces such as buses, electrolytic cells, and their attachments. The approach distances referred to in Table 130.4(E)(a) and Table 130.4(E)(b) shall not apply to work performed by qualified persons in the cell-line working zone.

Arc-Flash Risk Assessment

The requirements of 130.5, Arc-Flash Risk Assessment, shall not apply to electrolytic cell-line work zones. Each task performed in the electrolytic cell-line working zone shall be analyzed for the likelihood of arc-flash injury. If there is a likelihood of personal injury, appropriate measures shall be taken to protect persons exposed to the arc-flash hazards, including one or more of the following:

- Providing appropriate PPE to prevent injury from the arc-flash hazard

- Altering work procedures to reduce the likelihood of occurrence of an arc-flash incident

- Scheduling the task so that work can be performed when the cell line is de-energized

Before a routine task is performed in the cell line work zone, an arc-flash risk assessment shall be done. The results of the arc-flash risk assessment shall be used in training employees in job procedures that minimize the possibility of arc-flash hazards.

Before a nonroutine task is performed in the cell line working zone, an arc-flash risk assessment shall be done. If an arc-flash hazard is a possibility during nonroutine work, appropriate instructions shall be given to employees involved on how to minimize the risk associated with arc flash.

2. Batteries and Battery Rooms (Article 320)

Storage batteries are commonly used with uninterruptible power supplies (UPS), some types of power electronic equipment, and emergency control power for switching. Article 320 covers electrical safety requirements for the practical safeguarding of employees while working with exposed stationary storage batteries that exceed 50 volts, nominal.

Safety Requirements

- Prior to any work on a battery system, a risk assessment shall be performed to identify the chemical, electrical shock, and arc-flash hazards and assess the risks associated with the type of tasks to be performed.

- Article 320 requires specific warning signs for both chemical and electrical hazards.

- Tools and equipment for work on batteries shall be equipped with handles listed as insulated for the maximum working voltage.

- Battery terminals and all electrical conductors shall be kept clear of unintended contact with tools, test equipment, liquid containers, and other foreign objects.

⬛ SPECIAL ELECTRICAL EQUIPMENT

- Nonsparking tools shall be required when the risk assessment required by 110.5(H) justifies their use.

- Each battery room or battery enclosure shall be accessible only to authorized personnel.

- Employees shall not enter spaces containing batteries unless illumination is provided that enables the employees to perform the work safely.

- Do not wear electrically conductive objects such as jewelry while working on a battery system.

Battery Activities That Include Handling of Liquid Electrolyte

The following protective equipment shall be available to employees performing any type of service on a battery with liquid electrolyte:

1. Goggles and face shield appropriate for the electrical hazard and the chemical hazard.

2. Gloves and aprons appropriate for the chemical hazards.

3. Portable or stationary eye wash facilities and equipment within the work area that are capable of drenching or flushing of the eyes and body for the duration necessary to mitigate injury from the electrolyte hazard.

3. Lasers (Article 330)

A laser is defined as a device that produces radiant energy at wavelengths between 180 nm (nanometer) and 1 mm (millimeter) predominantly by controlled stimulated emission. Laser radiation can be highly coherent temporally, spatially, or both.

For the purpose of 330.3 in Article 330, hazardous voltage and current for AC systems is considered greater than or equal to 50 volts AC and 5 mA. For DC systems, hazardous voltage or current is considered greater than or equal to 100 volts DC and 40 mA. For the purpose of

🔌 SPECIAL ELECTRICAL EQUIPMENT

Article 330, hazardous stored energy is considered greater than or equal to 0.25 joules at 400 volts or greater, or 1 joule at greater than 100 volts up to 400 volts.

For work on or with lasers, training in electrical safe work practices shall include, but is not limited to, the following:

- Chapter 1 electrical safe work practices

- Electrical hazards associated with laser equipment

- Stored energy hazards, including capacitor bank explosion potential

- Ionizing radiation

- X-ray hazards from high-voltage equipment (>5 kV)

- Assessing the listing status of electrical equipment and the need for field evaluation of nonlisted equipment

The article does not describe the different laser classifications. Lasers are classified according to their capabilities of producing injury to personnel. These classifications range from Class I (no injury) to Class IV (able to cut thick steel). Class I lasers include levelers used on construction sites and laser pointers.

4. Power Electronic Equipment (Article 340)

This article discusses a range of industrial, medical, and communications gear, including:

- Electric arc welding equipment

- High-power radio, radar, and television transmitting towers and antennas

- Industrial dielectric and radio frequency (RF) induction heaters

- Shortwave or RF diathermy devices

SPECIAL ELECTRICAL EQUIPMENT

- Process equipment that includes rectifiers and inverters, such as motor drives, uninterruptible power supply systems, and lighting controllers

The employer and employees shall be aware of the hazards associated with the following:

- High voltages within the power supplies

- Radio frequency energy–induced high voltages

- Effects of RF fields in the vicinity of antennas and antenna transmission lines, which can introduce electrical shock and burns

- Ionizing (X-radiation) hazards from magnetrons, klystrons, thyratrons, cathode-ray tubes, and similar devices

Nonionizing RF radiation hazards from the following:

1. Radar equipment

2. Radio communication equipment, including broadcast transmitters

3. Satellite–earth transmitters

4. Industrial scientific and medical equipment

5. RF induction heaters and dielectric heaters

6. Industrial microwave heaters and diathermy radiators

Diathermy devices are medical equipment that use high-frequency radiation, microwaves, or ultrasound to heat and destroy abnormal cells.

5. Research and Development Laboratories (Article 350)

This article discusses requirements applicable to the electrical installations in areas, with custom or special electrical equipment,

SPECIAL ELECTRICAL EQUIPMENT

designated by the facility management for research and development (R&D) or as laboratories. The article establishes exceptions from Chapter 1 requirements that apply to tasks that are acceptable and necessary for special equipment used in R&D facilities but that might be unacceptable in routine industrial conditions.

R&D is defined as an activity in an installation specifically designated for research or development conducted with custom or special electrical equipment.

Each laboratory or R&D system application shall be permitted to assign an electrical safety authority (ESA) to ensure the use of appropriate electrical safety-related work practices and controls. The ESA shall be permitted to be an electrical safety committee, engineer, or equivalent qualified individual. The ESA shall be permitted to delegate authority to an individual or organization within their control.

The ESA shall act in a manner similar to an authority having jurisdiction for R&D electrical systems and electrical safe work practices. The ESA shall also be competent in the requirements of *NFPA 70E* and electrical system requirements applicable to the R&D laboratories.

Each laboratory or R&D system application shall be assigned a competent person (as defined in Article 350) to ensure the use of appropriate electrical safety-related work practices and controls.

A competent person is defined in Article 350 as a person who meets all the requirements of qualified person, as defined in *NFPA 70E,* and who, in addition, is responsible for all work activities or safety procedures related to custom or special equipment and has detailed knowledge regarding the exposure to electrical hazards, the appropriate control methods to reduce the risk associated with those hazards, and the implementation of those methods.

The equipment or systems used in the R&D area or in the laboratory shall be listed or field evaluated prior to use.

⎍ SPECIAL ELECTRICAL EQUIPMENT

6. Safety-Related Requirements for Capacitors (Article 360)

This article covers the electrical safety-related requirements for the practical safeguarding of employees while working with capacitors that present an electrical hazard.

Appropriate controls shall be applied where any of the following hazard thresholds are exceeded:

1. Less than 100 volts and greater than 100 joules of stored energy

2. Greater than or equal to 100 volts and greater than 1.0 joule of stored energy

3. Greater than or equal to 400 volts and greater than 0.25 joules of stored energy

The following qualifications and training shall be required for personnel safety:

1. Employees who perform work on electrical equipment with capacitors that exceed the energy thresholds in 360.3 shall be qualified and shall be trained in, and familiar with, the specific hazards and controls required for safe work.

2. Unqualified persons who perform work on electrical equipment with capacitors shall be trained in, and familiar with, any electrical safety-related work practices necessary for their safety.

The risk assessment process for capacitors shall follow the overall risk assessment procedures in NFPA 70E Chapter 1. If additional protective measures are required, they shall be selected and implemented according to the hierarchy of risk control identified in 110.5(H)(3). When the additional protective measures include the use of PPE, the following shall be determined:

1. Capacitor voltage and stored energy for the worker exposure.

2. Thermal hazard.

⚡ SPECIAL ELECTRICAL EQUIPMENT

3. Shock hazard.

4. Arc flash and arc blast hazard at the appropriate working distance.

5. Required test and grounding method.

6. Develop a written procedure that captures all of the required steps to place the equipment in an electrically safe work condition.

For more information on working safely with capacitors, see Informative Annex R, Working with Capacitors in NFPA 70E.

WHO ENFORCES COMPLIANCE WITH *NFPA 70E*?

Unlike the *NEC*, states, cities, and counties do not adopt *NFPA 70E* for regulatory purposes. State and municipal electrical inspectors do not enforce *NFPA 70E* unless the standard is adopted by local ordinance.

Instead, facility owners often require that both their employees and outside contractors comply with *NFPA 70E* while performing electrical construction and maintenance work. Increasingly, owners are requiring that contractors working for them provide evidence that their crews have been trained in *NFPA 70E* safety practices.

Doing so reduces the customers' liability exposure and insurance premiums.

NOTE: Retraining in safety-related work practices and applicable changes in NFPA 70E shall be performed at intervals not to exceed three years.

🔌 HOW IS *NFPA 70E* RELATED TO THE *NATIONAL ELECTRICAL CODE®*?

The *National Electrical Code (NEC)* describes design and installation of premise wiring systems, while *NFPA 70E* describes how to perform the work safely. Said another way, the *NEC* applies to equipment, circuits, and devices, while *NFPA 70E* applies to people. Both publications are also closely related to a third industry standard: *NFPA 70B, Recommended Practice for Electrical Maintenance.* Here is how the three work together:

• *NEC (NFPA 70)*—Describes how to design and install electrical systems that operate safely. It deals with subjects like overcurrent protection, conductor ampacity, wiring methods, equipment ratings, and grounding. The *NEC* does not cover maintenance of electrical equipment or safe work practice issues such as when workers should use PPE.

• *NFPA 70E*—Describes how to perform installation and maintenance work safely. It covers safe electrical work practices such as lockout/tagout, and when workers should use insulated tools and wear arc-rated clothing.

• *NFPA 70B*—Describes maintenance practices that keep electrical systems running reliably and safely. It does not cover design or installation issues and does not discuss safe work practices.

Knowledge Is Critical

Good electrical knowledge is needed to use *NFPA 70E* safely and effectively. *NFPA 70E* is organized in a way that is similar to the *NEC*, which helps electrical workers who are already familiar with the *NEC* to understand and apply the safety standard.

HOW IS *NFPA 70E* RELATED TO OSHA REGULATIONS?

All employers, in all industries, are legally required to follow OSHA regulations to protect their workers from job-related hazards. *NFPA 70E* is an American National Standard, developed by the publisher of the *NEC;* it parallels OSHA's electrical safety regulations.

However, *NFPA 70E* is not the *same* as OSHA regulations. It is more up to date than the OSHA electrical safety regulations. This guide summarizes the electrical safe work practices defined in *NFPA 70E*. For more information about how the standard is related to OSHA electrical regulations, see the Appendix and the following publication:

- *NFPA 70E: Handbook for Electrical Safety in the Workplace*

NFPA 70E is a private-sector, voluntary American National Standard that parallels the following OSHA safety regulations:

- OSHA 29 CFR 1910, General Industry Standards, Subpart S— Electrical (covers maintenance and repairs on existing systems).

- OSHA 29 CFR 1926, Construction Industry Standards, Subpart K— Electrical (covers new construction).

Most electrical contractors are required to follow both sets of OSHA safety regulations, depending on the type of work they are performing; 29 CFR 1926 covers *new* construction, before the facility is energized, whereas 29 CFR 1910 covers *maintenance* work. The two standards are similar but not quite matching, which can cause confusion.

OSHA Involvement

NFPA 70E was developed originally at OSHA's request. Because the federal government rule-making process is slow and cumbersome, keeping OSHA regulations up to date with current technology and work practices is difficult. For example, the OSHA 29 CFR 1926, Subpart K, regulations still reference the 1984 *NEC*.

HOW IS *NFPA 70E* RELATED TO OSHA REGULATIONS?

For this reason, private industry uses the *NFPA 70E* standard to "lead" the two sets of OSHA workplace safety regulations that govern electrical work: 1910 Subpart S for general industry applications, and 1926 Subpart K for the construction industry. OSHA personnel participate in the *NFPA 70E* Technical Committee.

NFPA 70E Enforcement

The practical result of complying with the safe work practices defined in *NFPA 70E* is in most cases complying also with the applicable OSHA regulations. Although OSHA safety and health compliance officers do not enforce *NFPA 70E* per se, there is a growing tendency for them to rely on *NFPA 70E* under the so-called "general duty" clause. Section 5(a)(1) of the Occupational Safety and Health Act requires employers to furnish safe workplaces that are free from recognized hazards that are causing or are likely to cause death or serious physical harm to employees.

WHO IS RESPONSIBLE FOR ELECTRICAL SAFETY?

NFPA 70E, like OSHA, states that both employers and employees are responsible for preventing injury.

- Employers shall provide safety-related work practices and shall train the employees.

- Employees shall implement the safety-related work practices established.

- Multiple employers often work together on the same construction site or in buildings and similar facilities. Some might be onsite personnel working for the host employer, while others are "outside" personnel such as electrical contractors, mechanical and plumbing contractors, painters, or cleaning crews. Outside personnel working for the host employer are employees of contract employers.

- *NFPA 70E* requires that when a host employer and contract employer work together within the limited approach boundary or the arc-flash boundary of exposed energized electrical conductors or circuit parts, they must coordinate their safety procedures.

- Where the host employer has knowledge of hazards covered by *NFPA 70E* that are related to the contract employer's work, there shall be a documented meeting between the host employer and the contract employer.

- Outside contractors are often required to follow the host employer's safety procedures.

- Multiple employers involved in the same project sometimes decide to follow the most stringent set of safety procedures.

- Whichever approach is taken, the decision should be recorded in the safety meeting documentation. In accordance with *NFPA 70E*, where the host employer has knowledge of hazards covered by *70E* that are related to the contract employer's work, there shall be a documented meeting between the host employer and the contract employer.

WHO IS RESPONSIBLE FOR ELECTRICAL SAFETY?

Who Is Responsible for PPE?

Both *NFPA 70E* and OSHA rules require various kinds of PPE, including insulated tools, face shields, and arc-rated clothing to protect electrical workers.

Employers are required to:

- Select appropriate PPE based on the hazards present or likely to be present in the workplace

- Prohibit the use of defective or damaged PPE

- Require that employees be trained so that each affected employee can properly use the assigned PPE

NOTE: Many Occupational Safety and Health Administration (OSHA) health, safety, maritime, and construction standards require employers to provide their employees with protective equipment, including personal protective equipment (PPE), when such equipment is necessary to protect employees from job-related injuries, illnesses, and fatalities. These requirements address PPE of many kinds: hard hats, gloves, goggles, safety glasses, welding helmets and goggles, face shields, chemical protective equipment, fall protection equipment, personal flotation devices (PFDs), and so forth. The provisions in OSHA standards that require PPE generally state that the employer is to provide such PPE. However, some of these provisions do not specify that the employer is to provide such PPE at no cost to the employee. In a rule that went into effect on May 15, 2008, OSHA now requires employers to pay for the PPE provided, with exceptions for specific items. The rule does not require employers to provide PPE where none has been required before. Instead, the rule merely stipulates that the employer must pay for required PPE, except in the limited cases specified in the standard.

Employees are responsible for implementing the safety-related work practices.

WHAT ARE ELECTRICAL HAZARDS?

Historically, shock and electrocution were seen as the primary electrical hazards to people, along with fires of electrical origin. Today, however, awareness of other electrical hazards is growing: arc flash, thermal flash, and arc blast. *NFPA 70E* defines four types of electrical hazards:

1. Electric shock

2. Arc-flash burn

3. Thermal burn

4. Arc blast injury

Electric Shock Hazard

- Electrical shock is the leading cause of death due to electricity.

- Tens of thousands of nonfatal electrical shock accidents and several hundred electrocutions (deaths from electrical shock) occur each year.

- More than half of the fatalities occur during the servicing of conductors and equipment energized at less than 600 volts.

- Shock occurs when a person contacts an exposed energized electrical conductor or circuit part or in some other way becomes part of an electrical circuit.

- Even the current needed to light a 7½-watt, 120-volt lamp (62.5 mA) is enough to kill a person if it passes across the chest and through the heart **(see Figure 42)**. *Note: Current (and time), not voltage, is the killer.*

FIGURE 42 Current flow through the human body. (Courtesy of Charles R. Miller)

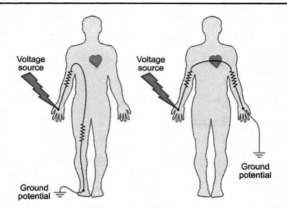

Ground-Fault Circuit Interrupter (GFCI) Protection

- *NFPA 70E*, *NEC*, and OSHA require GFCI protection on receptacles used to provide temporary power for construction and maintenance work. In accordance with *NFPA 70E*, GFCI protection shall be provided where an employee is operating or using cord sets (extension cords) or cord- and plug-connected tools related to maintenance and construction activity supplied by 125-volt, 15-, 20-, or 30-ampere circuits.

- GFCI circuit breakers and receptacles disconnect the circuit if fault current to ground (that might be traveling through a worker's body) exceeds 4–6 milliamperes (mA).

WHAT ARE ELECTRICAL HAZARDS?

Insulated Gloves and Tools

- Rubber insulating gloves, voltage-rated tools, and other insulating PPE are ways of protecting workers from electrical shock hazard.

- They are not intended for protection from arc-flash and arc-blast hazards.

Shock Approach Boundaries

- *NFPA 70E* defines two shock approach boundaries to energized electrical conductors and equipment: limited approach boundary and restricted approach boundary.

- These boundaries help protect workers against electrical shock hazard. *Note: Shock boundaries and arc-flash boundaries are not physically related.*

- Approach boundaries around energized electrical systems are explained in the earlier section of this guide under the heading "Safe Electrical Work Practices."

Arc-Flash Hazard

- When electrical current passes through air, temperatures can get hot enough to ignite clothing, melt plastic, and burn human skin. Arc flashes can and do kill at distances of up to 10 feet.

- Each year more than 2,000 people are admitted to burn centers with severe arc-flash burns.

- In addition to burn danger, the very bright light associated with arc-flash events can cause temporary (and sometimes permanent) blindness. An arc flash creates intense infrared (IR) and ultraviolet (UV) radiation; therefore, unprotected eyes can be damaged.

WHAT ARE ELECTRICAL HAZARDS?

Causes of Arc Flashes

- Although arc flashes can last longer, they are usually short events lasting less than a half-second (30 cycles).

- They can be caused by such things as dropping a tool or metal part across two energized busbars, operating malfunctioning or damaged switching equipment (so that an energized blade comes in contact with a metal case), or voltage testing with an improperly rated test instrument.

- Although an arc flash can be initiated by equipment failure, the vast majority of arc-flash events result from worker interaction.

Arc-Flash Boundary

- *NFPA 70E* defines an arc-flash boundary around energized electrical conductors and circuit parts to which an employee might be exposed.

- The boundary varies in size depending on the energy available in case of a fault.

- When an arc-flash hazard exists, the arc-flash boundary is an approach limit from an arc source at which incident energy equals 1.2 cal/cm^2.

Arc Flash and Voltage Levels

- Arc-flash hazard is related to the energy available in case of a fault.

- This energy level is related to the available short-circuit current and the clearing time of the circuit's overcurrent device (circuit breaker or fuse). The total clearing time for a fuse is the sum of the melt time plus arc time.

- Arc flash does not depend on system voltage alone. Under normal circumstances, 277- and 480-volt electrical systems present very serious arc-flash hazards.

WHAT ARE ELECTRICAL HAZARDS?

Arc-Blast Hazard

- Arc blast is a more severe version of arc flash. At higher available energy levels, explosion hazards—high pressure, flying shrapnel, and loud noise—are added to the thermal and light dangers of arc flashes. While arc flash is a thermal event, arc blast is a mechanical-pressure event.
- Extreme temperatures generated by the electric arc cause explosive expansion both of the surrounding air and of metal in the arc path. Copper expands many thousands of times when it vaporizes, creating a deadly spray of molten metal droplets.

Pressure Wave

This rapid expansion creates a pressure wave that can rupture equipment enclosures, knock workers off their feet, spray molten metal, and rupture eardrums.

Burn Dangers

- Oddly, workers exposed to arc blasts sometimes suffer fewer burns than those in arc-flash incidents.
- The blast energy can blow workers away from the arc, thus reducing their heat exposure but increasing the risk of serious falls, internal injuries, or impact injuries.

Contact Hazards and Noncontact Hazards

As these explanations show, the four electrical hazards defined in *NFPA 70E* actually fall into two general categories:

1. Contact hazards (shock and electrocution)
2. Noncontact hazards (arc-flash burn, thermal burn, and arc-blast injury)

Different techniques are used to protect workers against each category of hazard.

🔌 WHAT ARE ELECTRICAL HAZARDS?

Contact Hazards (Shock and Electrocution)

Shock or electrocution occurs when a person touches an energized (live) part and becomes part of a circuit.

NOTE: Thermal burn can also result from contact.

Qualified persons working on energized electrical systems can be protected against shock and electrocution by taking these precautions:

- Wearing rubber insulated gloves with leather protectors and other PPE

- Using insulated tools rated for the voltage

- Observing and complying with approach boundaries

Unqualified persons are protected against shock and electrocution by the following:

- Good maintenance practices (keeping electrical equipment in safe operating condition, making sure junction boxes have covers, ensuring that noncurrent-carrying metal parts of electrical equipment are grounded).

NOTE: Under normal operating conditions, enclosed energized equipment that has been properly installed and maintained is not likely to pose an arc-flash hazard.

- Barriers and warning signs

- Staying outside the limited approach boundary

WHAT ARE ELECTRICAL HAZARDS?

Noncontact Hazards (Arc-Flash Burn, Thermal Burn, and Arc-Blast Injury)

- Sometimes people are injured by being near an electrical failure.

- People can be injured by arc flashes and arc blasts even when they are not touching an exposed energized electrical conductor—sometimes when they are several feet away.

Qualified persons working on an energized electrical system can be protected against arc flash and arc blast by taking these precautions:

- Staying outside the arc-flash boundary will help protect against all burns. Nevertheless, there is still the possibility of a first-degree burn

- Wearing and using appropriate PPE, including arc-rated clothing, when they must work within the arc-flash boundary

NOTE: The first safety protocol is establishing an electrically safe work condition.

ELECTRICAL SAFETY PROGRAM

NFPA 70E requires that every employer have a written electrical safety program for work involving electrical systems or circuits operating at 50 volts or more. Informative Annex E of *NFPA 70E* describes general principles of such programs.

- The electrical safety program shall direct activity appropriate to the risk associated with electrical hazards.

- Include elements to verify that newly installed or modified electrical equipment or systems have been inspected to comply with applicable installation codes and standards prior to being placed into service.

- Include elements that consider the condition of maintenance of electrical equipment and systems.

- The program shall be designed to provide an awareness of the potential electrical hazards to employees who work in an environment with the presence of electrical hazards.

- The program shall be developed to provide the required self-discipline for all employees who must perform work that may involve electrical hazards.

- Ensure the electrical safety program instills safety principles and controls.

- The electrical safety program shall identify the principles upon which it is based.

- Identify the controls by which the program is measured and monitored.

- Identify the procedures to be utilized before work is started by employees exposed to an electrical hazard.

- The electrical safety program shall include a risk assessment procedure that shall: (1) address employee exposure to electrical

ELECTRICAL SAFETY PROGRAM

hazards; (2) identify the process to recognize hazards, assess risks; and (3) implement risk control according to the hierarchy of risk control methods.

The electrical safety program's risk assessment procedure shall address the potential for human error and its negative consequences on people, processes, work environment, and equipment relative to the electrical hazards in the workplace. The risk assessment procedure shall require that preventive and protective risk control methods be implemented in accordance with the following hierarchy: (1) elimination, (2) substitution, (3) engineering controls, (4) awareness, (5) administrative controls, and (6) personal protective equipment.

- The electrical safety program shall address job safety planning and job briefing.

- Ensure the program includes elements to investigate electrical incidents.

The electrical safety program shall be audited to verify that the principles and procedures of the electrical safety program are in compliance with *NFPA 70E*. Audits shall be performed at intervals not to exceed 3 years.

Field work shall be audited to verify that the requirements contained in the procedures of the electrical safety program are being followed. Audits shall be performed at intervals not to exceed 1 year.

The lockout/tagout program and procedures shall be audited by a qualified person at intervals not to exceed 1 year. The audit shall cover at least one lockout/tagout in progress.

TRAINING EMPLOYEES IN ELECTRICAL SAFETY

NFPA 70E requires that employers provide electrical safety training for employees who are exposed to an electrical hazard when the risk associated with that hazard is not reduced to a safe level. The training must include specific hazards associated with electrical energy. After the training, employees must be able to identify and understand the relationship between electrical hazards and possible injury associated with their respective job or task assignments.

Employees who must be trained include:

- Qualified persons

- Unqualified persons

The required training shall consist of:

- Classroom training

- On-the-job training

- A combination of the two

Although *NFPA 70E* does not specify the length of training (such as half-day, 1 day, 2 days, etc.), the degree of training shall be determined by the risk to the employee. It is also important to note that retraining in safety-related work practices and applicable changes in *NFPA 70E* shall be performed at intervals not to exceed 3 years.

Qualified Person

Because they face a greater risk of electrical hazards, qualified persons must be provided with more training than unqualified persons. Because equipment or a work method may present electrical hazards, qualified persons shall be knowledgeable in the construction and operation of equipment or a specific work method and shall be trained to identify and avoid those electrical hazards.

🔌 TRAINING EMPLOYEES IN ELECTRICAL SAFETY

NOTE: A person can be considered qualified with respect to certain equipment and methods but can remain unqualified with respect to other equipment and methods.

Training for qualified persons must also include proper use of:

- Special precautionary techniques
- Applicable electrical policies and procedures
- PPE
- Insulating and shielding materials
- Insulated tools
- Test equipment

Anyone permitted to work within the limited approach boundary shall, at a minimum, be additionally trained in all of the following:

- Skills and techniques necessary to distinguish exposed energized electrical conductors and circuit parts from other parts of electrical equipment
- Skills and techniques necessary to determine the nominal voltage of exposed energized electrical conductors and circuit parts
- Approach distances specified in Table 130.4(E)(a) and Table 130.4(E)(b) and the corresponding voltages to which the qualified person will be exposed
- Decision-making process is necessary to be able to do the following:
  1. Perform the job safety planning
  2. Identify electrical hazards
  3. Assess the associated risk
  4. Select the appropriate risk control methods from the hierarchy of controls identified in 110.5(H)(3), including PPE

TRAINING EMPLOYEES IN ELECTRICAL SAFETY

Training must include the selection process for an appropriate test instrument. Once trained, the employee must demonstrate how to use a voltage detector to verify the absence of voltage, including interpreting indications provided by the device. Because test instruments have limitations, the training shall include limitation information on each specific test instrument that might be used.

The employer shall determine through regular supervision or through inspections conducted on at least an annual basis that each employee is complying with the safety-related work practices required by *NFPA 70E*.

Unqualified Person

Unqualified persons, such as office workers, salespeople, cleaning crews, equipment operators, and workers in other construction trades who do not perform electrical work, shall be trained in and be familiar with any electrical safety-related practices necessary for their safety. However, unqualified persons must understand that they are not qualified to work on electrical equipment or circuits.

Retraining

Retraining in safety-related work practices and applicable changes in *NFPA 70E* shall be performed at intervals not to exceed 3 years. An employee shall receive additional training (or retraining) if any of the following conditions exist:

- The supervision or annual inspections indicate that the employee is not complying with safety-related work practices.

- New technology, new types of equipment, or changes in procedures necessitate the use of safety-related work practices that are different from those that the employee would normally use.

- The employee needs to review tasks that are performed less often than once per year.

⚡ TRAINING EMPLOYEES IN ELECTRICAL SAFETY

- The employee needs to review safety-related work practices not normally used by the employee during their regular job duties.

- The employee's job duties change.

Training Documentation

The employer shall document that each employee has received the training required by *NFPA 70E*.

This documentation shall be in accordance with the following:

- It shall be made when the employee demonstrates proficiency in the work practices involved.

- It shall be retained for the duration of the employee's employment.

- It shall contain the content of the training, each employee's name, and dates of training.

Content of the training could include one or more of the following: course syllabus, course curriculum, outline, table of contents, or training objectives.

Employment records that indicate that an employee has received the required training are an acceptable means of meeting this requirement.

Lockout/Tagout Procedure Training

Employees involved in or affected by the lockout/tagout procedures required by *NFPA 70E* 120.2(B) shall be trained in the following:

- The lockout/tagout procedures

- Their responsibility in the execution of the procedures

Retraining in the lockout/tagout procedures shall be performed as follows:

- When the procedures are revised

⚡ TRAINING EMPLOYEES IN ELECTRICAL SAFETY

- At intervals not to exceed 3 years

- When supervision or annual inspections indicate that the employee is not complying with the lockout/tagout procedures

Lockout/Tagout Training Documentation

- The employer shall document that each employee has received the training required by *NFPA 70E* 110.6(B).

- The documentation shall be made when the employee demonstrates proficiency in the work practices involved.

- The documentation shall contain the content of the training, each employee's name, and the dates of the training.

Content of the training could include one or more of the following: course syllabus, course curriculum, outline, table of contents, or training objectives.

Emergency Response Training

- Employees exposed to shock hazards and those responsible for the safe release of victims from contact with energized electrical conductors or circuit parts shall be trained in methods of safe release. Refresher training shall occur annually.

- Employees responsible for responding to medical emergencies shall be trained in first aid and emergency procedures.

- Employees responsible for responding to medical emergencies shall be trained in cardiopulmonary resuscitation (CPR).

- Employees responsible for responding to medical emergencies shall be trained in the use of an automated external defibrillator (AED) if an employer's emergency response plan includes the use of this device.

- Training shall occur at a frequency that satisfies the requirements of the certifying body.

TRAINING EMPLOYEES IN ELECTRICAL SAFETY

Employees responsible for responding to medical emergencies might not be first responders or medical professionals. Such employees could be a second person, a safety watch, or a craftsperson.

Emergency Response Training Verification and Documentation

- Employers shall verify at least annually that employee first aid, emergency response, and resuscitation training is current.

- The employer shall document that the employee first aid, emergency response, and resuscitation has occurred.

Job Safety Planning and Job Briefing

Before starting each job that involves exposure to electrical hazards, the employee in charge shall complete a job safety plan and conduct a job briefing with the employees involved.

The job safety plan shall be in accordance with the following:

- Be completed by a qualified person

- Be documented

- Include the following information:
  1. A description of the job and the individual tasks
  2. Identification of the electrical hazards associated with each task
  3. A shock risk assessment in accordance with 130.4 for tasks involving a shock hazard
  4. An arc-flash risk assessment in accordance with 130.5 for tasks involving an arc-flash hazard
  5. Work procedures involved, special precautions, and energy source controls

The job briefing shall cover the job safety plan and the information on the energized electrical work permit, if a permit is required.

TRAINING EMPLOYEES IN ELECTRICAL SAFETY

Additional job safety planning and job briefings shall be held if changes occur during the course of the work that might affect the safety of employees.

For an example of a job briefing form and planning checklist, see Informative Annex I, Figure I.1.

NFPA 70E has 18 informative annexes containing information that helps users comply with the standard. Brief summaries follow:

Informative Annex A—Informative Publications

This annex lists other industry standards referenced within *NFPA 70E* for informational purposes only and are thus not part of the requirements of *NFPA 70E*.

Informative Annex B—Reserved

This annex is reserved for the future.

Informative Annex C—Limits of Approach

This annex describes the two approach boundaries for shock protection: limited and restricted. This subject is covered on pages 38–42.

Informative Annex D—Incident Energy and Arc Flash Boundary Calculation Methods

This annex summarizes calculation methods available for calculating arc-flash boundary and incident energy. Table D.1 summarizes the limitations for the calculation methods described. The arc-flash boundary for systems 50 volts and greater shall be the distance at which the incident energy equals 1.2 cal/cm².

TRAINING EMPLOYEES IN ELECTRICAL SAFETY

Informative Annex E—Electrical Safety Program

This annex summarizes basic principles of company electrical safety programs. It provides examples of typical electrical safety program principles, controls, and procedures.

Informative Annex F—Risk Assessment and Risk Control

NFPA 70E requires a risk assessment procedure included as part of the electrical safety program. This annex provides guidance regarding a qualitative approach for risk assessment, which can be helpful in determining the protective measures that are required to reduce the likelihood of injury or damage to health occurring under the circumstances being considered. This annex explains the hierarchy of risk control methods and shows an example of a qualitative two-by-two risk assessment matrix.

Informative Annex G—Sample Lockout/Tagout Program

This annex contains a sample company lockout–tagout program. This annex shows both simple lockout/tagout procedures and complex lockout/tagout procedures. Page 9 of this guide describes the procedure for establishing an electrically safe work condition.

Informative Annex H—Guidance on Selection of Protective Clothing and Other Personal Protective Equipment (PPE)

This annex provides guidance in selecting arc-rated clothing and other PPE. It focuses on two tables. The first table is a simplified approach to provide minimum PPE for electrical workers within facilities with large and diverse electrical systems. The second table provides a summary of specific sections within the *NFPA 70E* standard describing PPE for electrical hazards.

TRAINING EMPLOYEES IN ELECTRICAL SAFETY

Informative Annex I—Job Briefing and Job Safety Planning Checklist

This annex contains a sample checklist for job briefings to plan electrical work and a checklist for job safety planning. Page 141 of this *Ugly's* guide discusses job safety planning and job briefings.

Informative Annex J—Energized Electrical Work Permit

This annex contains a sample Energized Electrical Work Permit and an Energized Electrical Work Permit Flowchart. The energized electrical work permit flow chart is extremely helpful in determining whether an energized electrical work permit is needed. Page 34 of this *Ugly's* guide discusses an Energized Electrical Work Permit.

Informative Annex K—General Categories of Electrical Hazards

This annex describes three general categories of electrical hazards: shock, arc flash, and arc blast.

Informative Annex L—Typical Application of Safeguards in the Cell Line Working Zone

This annex describes safeguards for working around electrolytic cell lines covered by *NFPA 70E*, Article 310.

Informative Annex M—Layering of Protective Clothing and Total System Arc Rating

This annex discusses issues associated with the protective nature of multiple layers of arc-rated clothing. Layering of arc-rated clothing is an effective approach to achieving the required arc rating when the layers have been tested together to determine the composite rating.

Informative Annex N—Example Industrial Procedures and Policies for Working Near Overhead Lines and Equipment

This informative annex is an example of an industrial procedure for working near overhead electrical systems. Areas covered include operations that could expose employees or equipment to contact with overhead electrical systems.

TRAINING EMPLOYEES IN ELECTRICAL SAFETY

Informative Annex O—Safety-Related Design Requirements

This annex illustrates that many decisions made during the design of a facility have an important bearing on safe work practices.

Informative Annex P—Aligning Implementation of This Standard with Occupational Health and Safety Management Standards

This annex provides additional information on how *NFPA 70E* aligns with the framework promoted in ANSI-AIHA Z10 *American National Standard for Occupational Health and Safety Management Systems* and other internationally recognized Occupational Health and Safety Management System standards.

Informative Annex Q—Human Performance and Workplace Electrical Safety

This informative annex introduces the concept of human performance and how this concept can be applied to workplace electrical safety. The objective of human performance is to identify and address human error and its negative consequences on people, programs, processes, the work environment, an organization, or equipment.

Informative Annex R—Working with Capacitors

Article 360 addresses specific electrical safety requirements unique to capacitors. This annex contains nonmandatory information relative to working with capacitors. Capacitors have the ability to store electrical energy after the source power has been disconnected.

Conclusion

Several of the annexes to *NFPA 70E*, particularly Annex H, Annex I, and Annex J, contain useful technical information that is critical to correct application of the standard. The remaining annexes provide general illustrations or administrative information. Informative annexes are not part of the requirements of NFPA 70E but are included for information purposes only.

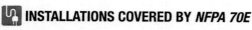

INSTALLATIONS COVERED BY *NFPA 70E*

For complete information about which electrical installations are covered or not covered, see *NFPA 70E*. The following lists of **covered** and **not covered** installations provide useful examples.

Covered

NFPA 70E applies to electrical installations operating at 50 volts and above, on the customer side of the service point. Examples include, but are not limited to, the following:

- 120/240-volt, single-phase, 3-wire power systems

- 208Y/120-volt, 3-phase, 4-wire power systems

- 480Y/277-volt, 3-phase, 4-wire power systems

- 480-volt, 3-phase, 3-wire systems

- Class 1 control wiring operating at more than 50 volts (covered by *NEC* Article 725)

- 120-volt, motor-control wiring

- 600Y/347-volt, 3-phase, 4-wire power systems (*NOTE: Although this system voltage is common in Canada, it is rarely used in the United States.*)

- Privately owned power systems operating at above 600 volts. These include installations such as industrial power systems operating at 4160Y/2400 volts and campus-wide power distribution systems operating at 7200 volts or more.

Not Covered

NFPA 70E does not apply to electrical installations operating at less than 50 volts in which an increased exposure to electrical burns or to explosion due to electric arcs is not present. Increased exposure is determined through consideration of the capacity of the energy

INSTALLATIONS COVERED BY *NFPA 70E*

source and any overcurrent protection between the source and the worker. These installations include most signaling, communications, and control installations, along with low-voltage lighting. Examples of systems not covered by *NFPA 70E* include the following:

- Telecommunications systems operating at 48 volts

- Fire alarm systems operating at 24 volts and 32 volts

- Class 2 and 3 wiring systems (covered by *NEC* Article 725)

- Remote control systems such as 12-volt lighting control

- Optical fiber communications systems (*NOTE: Composite cables containing both optical fibers and electric power conductors operating at 50 volts or higher are covered by* NFPA 70E.)

- Low-voltage lighting systems operating at a maximum of 30 volts (covered by *NEC* Article 411)

- Utility generating, transmission, and distribution systems

- DC power systems for railroads, subways, and similar purposes

- Airplanes, watercraft, and automotive vehicles (other than recreational vehicles)

NFPA 70E DEFINITIONS

NOTE: Some NFPA 70E definitions include informational notes, which are shown below. Comments shown in italics under some definitions are additional explanations that do not appear in the NFPA 70E standard. Some definitions not covered in this section are covered in other parts of this book.

Arc-Flash Hazard: A source of possible injury or damage to health associated with the possible release of energy caused by an electric arc.

Informational Note No. 1: The likelihood of occurrence of an arc flash incident increases when energized electrical conductors or circuit parts are exposed or when they are within equipment in a guarded or enclosed condition, provided a person is interacting with the equipment in such a manner that could cause an electric arc. An arc-flash incident is not likely to occur under normal operating conditions when enclosed energized equipment has been properly installed and maintained.

Informational Note No. 2: See Table 130.5(C) for examples of tasks that increase the likelihood of an arc-flash incident occurring.

Arc-Flash Boundary: When an arc-flash hazard exists, an approach limit from an arc source at which incident energy equals 1.2 cal/cm² (5 J/cm²).

Informational Note: According to the Stoll skin burn injury model, the onset of a second-degree burn on unprotected skin is likely to occur at an exposure of 1.2 cal/cm² (5 J/cm²) for 1 second.

Comment: It is important to understand that staying outside the arc-flash boundary does not guarantee that a person will not be burned by an arc-flash incident. It means the person could receive a curable burn rather than a more serious or fatal burn.

Arc-Flash Suit: A complete arc-rated clothing and equipment system that covers the entire body except for the hands and feet.

NFPA 70E **DEFINITIONS**

Equipment, Arc-Resistant: Equipment designed to withstand the effects of an internal arcing fault and direct the internally released energy away from the employee.

Comment: Equipment, Arc-resistant is just one design available to minimize energy levels where employees are exposed to electrical hazards. Other equipment and designs include, but are not limited to, remote racking (insertion or removal), remote opening and closing of switching devices, high-resistance grounding of low-voltage and 5 kV (nominal) systems, current limitation, and specification of covered bus or covered conductors within equipment.

De-energized: Free from any electrical connection to a source of potential difference and from electrical charge; not having a potential different from that of the earth.

Comment: This is a key concept of NFPA 70E. *The safest way to work on electrical conductors and equipment is de-energized. See* Electrically Safe Work Condition.

Disconnecting Means: A device, or group of devices, or other means by which the conductors of a circuit can be disconnected from their source of supply.

Comments:

• *Many different devices are permitted to serve as required disconnecting means for different types of equipment. These include wall switches and attachment plugs, under certain circumstances.*

• *In accordance with 29 CFR 1910.303(f)(4), disconnecting means required by 1910 Subpart S—Electrical shall be capable of being locked in the open position.*

• *After January 2, 1990, whenever replacement or major repair, renovation, or modification of a machine or equipment is performed, and whenever new machines or equipment are installed, energy*

NFPA 70E DEFINITIONS

isolating devices for such machine or equipment shall be designed to accept a lockout device [29 CFR 1910.147(c) (2)(iii)].

- *The common field terms "disconnect," "disconnect switch," and "safety switch" are not used in NFPA 70E (or the NEC).*

- *Circuit breakers can be used as disconnecting means. Fuses by themselves are not considered disconnecting means. However, a fused disconnect switch or pullout block that simultaneously disconnects all ungrounded (phase) conductors of a circuit is considered a disconnecting means.*

Electrical Hazard: A dangerous condition such that contact or equipment failure can result in electrical shock, arc-flash burn, thermal burn, or arc-blast injury.

Informational Note: Class 2 power supplies, listed low-voltage lighting systems, and similar sources are examples of circuits or systems that are not considered an electrical hazard.

Comment: NFPA 70E safe work practices apply to systems energized at 50 volts and above, both AC and DC.

Electrical Safety: Identifying hazards associated with the use of electrical energy and taking precautions to reduce the risk associated with those hazards.

Electrical Safety Program: A documented system consisting of electrical safety principles, policies, procedures, and processes that directs activities appropriate for the risk associated with electrical hazards.

Electrically Safe Work Condition: A state in which an electrical conductor or circuit part to be worked on or near has been disconnected from energized parts, locked/tagged in accordance with established standards, tested to verify the absence of voltage, and, if necessary, temporarily grounded for personnel protection.

📋 NFPA 70E DEFINITIONS

Informational Note: An electrically safe work condition is not a procedure, it is a state wherein all hazardous electrical conductors or circuit parts to which a worker might be exposed are maintained in a de-energized state for the purpose of temporarily eliminating electrical hazards for the period of time for which the state is maintained.

Comment: This is a key concept of NFPA 70E. The safest way to work on electrical conductors and equipment is de-energized. The process of turning off the electricity, verifying that it is off, and ensuring that it stays off while work is performed is called "establishing an electrically safe work condition." Many people call the process of ensuring that the current is removed "lockout/tagout"; however, lockout/tagout is only one step in the process.

Energized: Electrically connected to, or is, a source of voltage.

Fault Current: The amount of current delivered at a point on the system during a short-circuit condition.

Fault Current, Available: The largest amount of current capable of being delivered at a point on the system during a short-circuit condition.

Informational Note No. 1: A short circuit can occur during abnormal conditions such as a fault between circuit conductors or a ground fault.

Informational Note No. 2: If the DC supply is a battery system, the term "available fault current" refers to the prospective short-circuit current.

Informational Note No. 3: The available fault current varies at different locations within the system due to the location of sources and system impedances.

Hazard: A source of possible injury or damage to health.

Hazardous: Involving exposure to at least one hazard.

Incident Energy: The amount of thermal energy impressed on a surface, a certain distance from the source, generated during an electrical arc event. Incident energy is typically expressed in calories per square centimeter (cal/cm^2).

 **NFPA 70E DEFINITIONS**

Incident Energy Analysis: A component of an arc-flash risk assessment used to predict the incident energy of an arc flash for a specified set of conditions.

Limited Approach Boundary: An approach limit at a distance from an exposed energized electrical conductor or circuit part within which a shock hazard exists.

Comment: This term describes a distance from exposed energized electrical conductors or circuit parts beyond which the risk of electrical shock is considered low. Workers at least this far away from the equipment do not have to take special safety precautions. [See Restricted Approach Boundary, *and* Working On (energized electrical conductors and circuit parts)]. *Approach boundaries are related only to shock exposure; they have nothing to do with arc-flash protection. The distance for the arc-flash boundary could be greater or less than the limited approach boundary.*

Premises Wiring (System): Interior and exterior wiring, including power, lighting, control, and signal circuit wiring, together with all their associated hardware, fittings, and wiring devices, both permanently and temporarily installed. This includes (a) wiring from the service point or power source to the outlets; or (b) wiring from and including the power source to the outlets where there is no service point. Such wiring does not include wiring internal to appliances, luminaires, motors, controllers, motor control centers, and similar equipment.

Comment: As in the NEC, *this general term covers all power, communications, and control wiring of a building or similar structure from the service point, point of entry, or other source to the outlets.*

Qualified Person: One who has demonstrated skills and knowledge related to the construction and operation of electrical equipment and installations and has received safety training to identify the hazards and reduce the associated risk.

Comment: A person can be considered qualified with respect to certain equipment and methods but still be unqualified for others.

NFPA 70E DEFINITIONS

Restricted Approach Boundary: An approach limit at a distance from an exposed energized electrical conductor or circuit part within which there is an increased likelihood of electric shock due to electrical arc-over combined with inadvertent movement.

Comment: See Limited Approach Boundary, *and* Working On (exposed energized electrical conductors or circuit parts). *Approach boundaries are related only to shock exposure; they have nothing to do with arc-flash protection.*

Risk: A combination of the likelihood of occurrence of injury or damage to health and the severity of injury or damage to health that results from a hazard.

Risk Assessment: An overall process that identifies hazards, estimates the likelihood of occurrence of injury or damage to health, estimates the potential severity of injury or damage to health, and determines if protective measures are required.

Informational Note: As used in *NFPA 70E*, arc-flash risk assessment and shock risk assessment are types of risk assessments.

Service Point: The point of connection between the facilities of the serving utility and the premises wiring.

Comment: This definition is a crucial concept in both NFPA 70E *and the* NEC. *Conductors and equipment on the customer side (load side—downstream) of the service point are covered by* NFPA 70E *and* NEC *rules. The* NEC *does not cover conductors and equipment on the utility side (line side—supply side—upstream) of the service point. Typically, these are constructed according to the* National Electrical Safety Code (NESC) *or the serving utility's own rules.*

Shock Hazard: A source of possible injury or damage to health associated with current through the body caused by contact or approach to exposed energized electrical conductors, or circuit parts.

 **NFPA 70E DEFINITIONS**

Informational Note: Injury and damage to health resulting from shock is dependent on the magnitude of the electrical current, the power source frequency (e.g., 60 Hz, 50 Hz, DC), and the path and time duration of current through the body. The physiological reaction ranges from perception to muscular contractions, to inability to let go, to ventricular fibrillation, to tissue burns, to death.

Short-Circuit Current Rating: The prospective symmetrical fault current at a nominal voltage to which an apparatus or system is able to be connected without sustaining damage exceeding defined acceptance criteria.

Single-Line Diagram: A diagram that shows, by means of single lines and graphic symbols, the course of an electrical circuit or system of circuits and the component devices or parts used in the circuit or system.

Step Potential: A ground potential gradient difference that can cause current flow from foot to foot through the body.

Comment: See Touch Potential.

Switch, Isolating: A switch intended for isolating an electrical circuit from the source of power. It has no interrupting rating, and it is intended to be operated only after the circuit has been opened by some other means.

Touch Potential: A ground potential gradient difference that can cause current flow from hand to hand, hand to foot, or another path other than foot to foot, through the body.

Comment: See Step Potential.

Unqualified Person: A person who is not a qualified person.

Comment: Unqualified persons are those not trained in the construction, operation, and safety aspects of electrical equipment and systems. See Qualified Person.

NFPA 70E DEFINITIONS

Working Distance: The distance between a person's face and chest area and a prospective arc source.

Informational Note: Incident energy increases as the distance from the arc source decreases.

Working On (energized electrical conductors and circuit parts): Intentionally coming in contact with energized electrical conductors or circuit parts with the hands, feet, or other body parts; with tools, probes, or test equipment, regardless of the personal protective equipment (PPE) a person is wearing. There are two categories of "working on." *Diagnostic (testing)* is taking readings or measurements of electrical equipment, conductors, or circuit parts with approved test equipment that does not require making any physical change to the electrical equipment, conductors, or circuit parts. Repair is any physical alteration of electrical equipment, conductors, or circuit parts (such as making or tightening connections, removing or replacing components).

⏻ GENERAL PROTECTION FROM ELECTRICAL INJURIES

Most people think of PPE only as clothing or rubber insulating products. This text has already discussed PPE that guards against the hazards associated with shock and arc flash, even though some of these items might not normally be thought of as PPE. This section considers general PPE that can help prevent injuries against those and other hazards in electrical work. The following topics are important for safe electrical practices:

• Hard hats

• Safety glasses and safety goggles

• Face shields and viewing windows

• Voltage-rated hand tools

• Temporary protective grounding equipment

Hard Hats

Hard hats are intended to provide head protection from falling objects and bumping, and to prevent the head from contacting energized conductors **(see Figure 43)**. Gravity usually causes falling objects to travel straight down from an elevated position. The forces of falling objects are applied directly to the top of the protective helmets and distributed onto the worker's head by the helmet's suspension system. Sometimes a falling object strikes a fixed object or structure and is directed into the worker's head from the side. The helmet assembly must distribute the kinetic energy contained in the falling object in this instance as well.

GENERAL PROTECTION FROM ELECTRICAL INJURIES

FIGURE 43   Protective helmets (hard hats). (Courtesy of Tasco-safety.com)

The test setup for the different types of exposure to the hazards of falling objects or bumping must account for the various directions of the forces applied to the hard hat. National consensus standards use types to differentiate these exposures:

- Type I helmets are tested to distribute downward forces adequately.

- Type II helmets are tested to distribute both downward and lateral forces adequately.

Suspension systems may have four or six points of support for the hard hat **(see Table 9)**. Six-point systems provide greater distribution of the impact.

GENERAL PROTECTION FROM ELECTRICAL INJURIES

TABLE 9   Classes of Hard Hats

Class	Maximum Voltage Rating	Equivalent Older Designation
E	20 kV	B
G	2.2 kV	A
C	No rating	C

Note: Class C hard hats are conductive.
Reproduced from ANSI/ISEA Z89.1-2014

ANSI Z89.1 *Requirements*

One main ANSI national consensus standard now offers guidelines for head protection: *ANSI/ISEA Z89.1, American National Standard for Industrial Head Protection.*

The standard includes minimum performance requirements for industrial head protection that is designated by both its Type (location of impact and penetration) and Class (electrical insulation). In 1997, the electrical insulation designations of Class A, B, and C were changed, respectively, to Class E (electrical), G (general), and C (conductive) to make the kind of protection offered by the helmet more intuitive to the user.

OSHA Requirements

In *29 CFR 1910.132*, OSHA requires employers to ensure that workers wear hard hats that provide protection from falling objects.

- Employers must execute and document a hazard analysis to decide what type of head protection workers should use.

- If the potential injury is limited to a worker bumping their head on an obstruction, the head protection need only consist of bump caps, which are lightweight and less restricting than heavier hard hats.

GENERAL PROTECTION FROM ELECTRICAL INJURIES

- However, use of bump caps is restricted to applications where falling objects are not anticipated.

A requirement for protective helmets that reduce the electrical shock hazard is cited in *29 CFR 1910.135(a)(2)*. In *1910.268(j)(1)*, Class B head protection is required when a worker might be exposed to high-voltage electrical contact. [Note that "Class B" was a designation in the now superseded *ANSI Z89.2*. The current *ANSI Z89.1* calls this head protection Class E.] In *1910.335(a)(1)(4)*, OSHA requires that workers wear head protection where danger from shock or burns is possible. The employer is responsible for ensuring that employees comply with these requirements.

ANSI Z89.1-2014 assigns helmets to a class as defined by voltage. Note that Class C hard hats are conductive and should not be worn by workers who are or may be exposed to an energized, uninsulated electrical conductor.

Selection and Use

- Hard hats are available in many different colors.

- Some employers provide workers in each discipline or craft with a unique color.

- In other instances, the color of the worker's hard hat indicates contract workers or specific contractors.

- In either instance, the hard hat should provide the impact protection and electrical shock protection defined in *ANSI Z89.1-2014*. The best alternative is for all workers to wear head protection rated as Class E protective hard hats.

- Hard hats are made from various moldable materials. Polyethylene is common, because it can be molded easily into the required form and offers excellent resistance to abrasion and breaking. Hard hats may be constructed from other materials, provided the completed assembly meets the impact, penetration, and insulating characteristics defined in *ANSI Z89.1*.

- Current national consensus standards for protective helmets define tests for impact, penetration, and electrical conductivity.

- However, ignition and flammability characteristics for protection from arc-flash events are not addressed.

- Although ignition and flammability are important for use by fire fighters, exposure to an arc-flash event is not currently considered important for construction of hard hats.

- A hard hat made from polyethylene and similar materials could ignite when exposed to an arcing fault and should be covered by an arc-rated hood or similar product if the arc-flash risk assessment indicates that the worker may be exposed to an arc-flash event.

- Helmets marked with a "reverse donning arrow" can be worn frontward or backward in accordance with the manufacturer's instructions. They pass all testing requirements, whether worn frontward or backward.

- Workers sometimes put decals on the outer surface of their hard hat. If the hard hat is Class E or G (i.e., rated for use near exposed, energized electrical conductors), the decal should be nonconductive. Metal or conductive decals could be the cause of a short circuit and initiate an arcing fault.

- Chinstraps and other attachments are available for hard hats. Adding an attachment might provide additional utility; however, the attachment might increase the potential for damage from a different hazard. For instance, a chinstrap will prevent the helmet from falling off the head, but can increase the amount of fuel if it should be ignited in an arc-flash event.

GENERAL PROTECTION FROM ELECTRICAL INJURIES

Purchase specifications for hard hats must identify the characteristics of the desired helmet. They must indicate size (small, standard, or large), type (Type I or Type II), and color. When a hard hat is purchased, the suspension system also must be purchased. The suspension system must match the type and size of the hard hat. Incorrect application of the suspension system negates the protective characteristics of the protective helmet. Suspension systems and protective helmets from different manufacturers must not be intermingled. Hard hats are sold through distributors and local safety equipment suppliers.

Safety Glasses and Safety Goggles

- National consensus standards define performance criteria, testing requirements, and required marking for safety glasses. *ANSI Z87.1, American National Standard for Occupational and Educational Personal Eye and Face Protection Devices*, indicates that safety glasses can be either nonimpact or impact protectors.

- Products marked as impact protectors must pass all high-impact testing requirements and will be marked as "Z87+." Non-impact protectors are those that do not pass all high-impact testing requirements and are marked only with "Z87" (no "+" sign).

- The terms "eye glasses" and "safety glasses" are commonly used to mean an assembly of lenses together with a supporting frame that is worn to correct a person's vision or protect a person's eyes from impact **(see Figure 44)**.

- The term "spectacles" is used in consensus and regulatory standards to have the same meaning. Spectacles, however, excludes coverall safety glasses and goggles **(see Figure 45)**.

In *29 CFR 1910.133(b)(2)*, OSHA suggests that protective equipment for eyes must comply with the requirements of *ANSI Z87.1*. This standard defines performance requirements that guide the construction

FIGURE 44   Safety glasses. (Courtesy of Oberon Company)

of frames and lenses. By setting performance requirements, frames and lenses and support structures may be made from several different materials and with different construction, provided the overall assembly complies with the specified performance criteria.

Safety glasses, as such, are not electrical PPE. However, safety glasses could fall from a worker's face and initiate an arcing fault if the protective equipment is conductive. In some instances, the supporting frame for safety lenses is made of metal and is conductive. Unrestrained spectacles (safety glasses or otherwise) must not be worn when a worker is performing work on or near exposed live parts.

In accordance with *NFPA 70E*, employees shall wear protective equipment for the eyes whenever there is danger of injury from electric arcs, flashes, or from flying objects resulting from electrical explosion. When wearing an arc-rated flash suit hood or an arc-rated hood, eye protection (safety glasses or goggles) shall always be worn under the hood or face shield.

GENERAL PROTECTION FROM ELECTRICAL INJURIES

FIGURE 45   Safety goggle and safety spectacle. (Courtesy of Oberon Company)

Face Shields and Viewing Windows

Every work task should begin with an arc-flash risk assessment. Workers cannot select and wear adequate PPE until and unless all potential hazards have been identified. Unless workers are familiar with the nature and limits of protective characteristics for available PPE, the workers cannot choose the protective equipment with confidence. Workers, supervisors, and managers must be trained to recognize the protective limits provided by equipment that meets national standards.

GENERAL PROTECTION FROM ELECTRICAL INJURIES

ANSI Z87.1 *Requirements*

Face shields can provide protection from several hazards.

- In some instances, a face shield is necessary to provide protection from impact. Grinding and cutting tasks usually generate flying objects that could injure a worker's face or eyes.

- A face shield that meets the impact and penetration requirements defined in *ANSI Z87.1, American National Standard for Occupational and Educational Personal Eye and Face Protection Devices*, will provide adequate protection in this situation.

- When workers are required to work with and handle liquid chemicals, a face shield will provide protection from occasional splashes that might occur. If the face shield is impervious to attack from the chemical, the face shield will provide adequate protection. If the face shield used for protection from a chemical spill also meets the impact requirements of *ANSI Z87.1*, the same face shield may be worn for protection from both hazards.

- The same face shield that provides chemical and impact protection will filter some wavelengths of energy in the electromagnetic spectrum.

- Face shields that meet the requirements of *ANSI Z87.1* provide protection from nonionizing radiated energy. Some ultraviolet energy is filtered as well. However, face shields and other equipment covered by *ANSI Z87.1* do not provide protection from infrared energy or thermal energy **(see Figure 46)**.

An arcing fault generates significant thermal energy in the form of heated air and gases that rush toward the worker's face at speeds approaching the speed of sound. Molten copper and parts and pieces of the equipment might also be rushing toward the worker's face at near the speed of sound. Significant electromagnetic energy is

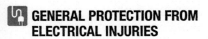

GENERAL PROTECTION FROM ELECTRICAL INJURIES

generated in an arcing fault and is radiated toward the worker's face at the speed of light. Workers should wear equipment that provides protection from all of these destructive components. An ordinary face shield intended for protection when using a grinder will not protect a worker from exposure to hazards associated with an arcing fault.

FIGURE 46   Arc-rated face shield. (Courtesy of Oberon Company)

Direct contact between an exposed energized conductor or circuit part and a worker's face is unlikely. However, a worker's face is very likely to be exposed to the heated gases and molten metal associated with an arcing fault.

- The worker's face and head must be protected from the destructive nature of these hazards.

- Therefore, the face shield must both meet impact requirements and provide adequate protection from the thermal hazard.

GENERAL PROTECTION FROM ELECTRICAL INJURIES

- Eye protection (safety glasses or goggles) shall always be worn under a face shield.

ASTM F2178 *Requirements*

ASTM F2178, Standard Specification for Arc Rated Eye or Face Protective Products, defines a testing method to establish a rating for face shields and viewing windows where the product will be exposed to an arcing fault.

- The rating system established in this standard suggests that products should be assigned an arc rating that essentially mirrors ratings for clothing defined in *ASTM F1506 Standard Performance Specification for Flame Resistant and Electric Arc Rated Protective Clothing Worn by Workers Exposed to Flames and Electric Arcs.*

- The arc rating should be based on the arc thermal performance value (ATPV) or the energy of breakopen threshold (E_{BT}). Manufacturers assign arc ratings based on internal tests. No third-party testing process exists for establishing an arc rating.

- Face shield selection should be based on the expected available incident energy that is identified when the arc-flash risk assessment is performed.

- A face shield that has a rating equal to or greater than the expected incident energy exposure shall be selected.

- When face protection is an integral part of another item of protective apparel, the viewing window should at least have the same rating as the other protective apparel. Consult the manufacturer of the protective apparel.

- A face shield provides an edge on each side of the head of the person wearing it.

GENERAL PROTECTION FROM ELECTRICAL INJURIES

- A face shield obstructs the movement of the air and gases rushing from the arcing fault during an exposure.

- Objects such as molten metal will be stopped by the obstructing nature of the face shield. However, the obstructing nature and curved shape of a face shield also obstructs the movement of air and gases rushing from an arcing fault. If the worker's head is positioned such that the rushing air is not directed at the front of the face shield, the curvature of the face shield could result in a vacuum being generated behind the face shield by the air and gases rushing past. Lift is generated under an airplane wing by the same action **(see Figure 47)**.

FIGURE 47   Illustration of lift.

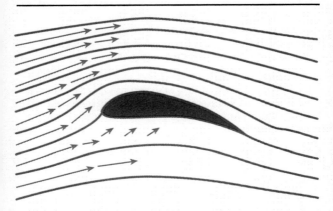

All parts of a worker's body that are within the arc-flash boundary must be protected from the effects of the potential arcing fault. If the worker's entire head is within the arc-flash boundary, a face shield will not provide adequate protection. An arc-rated balaclava shall be used with an arc-rated face shield when the back of the head is within the arc-flash boundary. An arc-rated hood shall be used instead of an arc-rated face shield and balaclava when the anticipated incident energy exposure exceeds 12 cal/cm².

Voltage-Rated Hand Tools

In *29 CFR 1910.333(c)(2)*, OSHA suggests that voltage-rated tools must be used for all work where the hand tools might make contact with an exposed energized conductor. *NFPA 70E* also contains a similar requirement. *ASTM F1505, Standard Specification for Insulated and Insulating Hand Tools*, defines construction and testing of hand tools that are rated for use on circuits that are 1,000 Vac (volts of alternating current) or 1,500 Vdc (volts of direct current).

- The term *voltage-rated hand tools* refers to both insulated and insulating hand tools. Insulated hand tools are constructed from conductive material or components and have electrical insulation applied on the exterior surface **(see Figure 48)**.

- In accordance with *NFPA 70E*, employees shall use insulated tools or handling equipment, or both, when working inside the restricted approach boundary of exposed energized electrical conductors or circuit parts where tools or handling equipment might make unintentional contact.

- Voltage-rated hand tools are not intended to serve as primary protection from shock or electrocution.

- Although insulated hand tools might provide adequate shock protection for circuits below 1,000 Vac, workers must select and wear PPE that provides protection from shock and arc flash without considering the hand-tool rating.

- Although insulated and insulating hand tools include insulation from electrical sources of up to 1,000 Vac, the primary function of the insulation is to reduce the risk of initiating an arcing fault.

GENERAL PROTECTION FROM ELECTRICAL INJURIES

FIGURE 48   Voltage-rated hand tools. (Courtesy of Oberon Company)

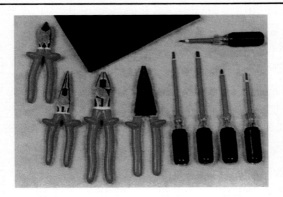

- Insulated tools shall be protected from damage to the insulating material.

- The insulating coating of insulated hand tools may consist of a single layer or two layers of contrasting colors.

- The interior layer may or may not provide 100% protection from shock to the full rating of the tool; check with the manufacturer.

- The contrasting color of the exterior layer provides a method for inspecting the tool for damage.

- Any cut or abrasion that exposes any of the inner layers constitutes significant damage and suggests that the tool should not be used. The damaged tool should be replaced with a new tool.

- Visually inspect each voltage-rated tool before each use. The inspection should involve looking for damage to the insulation

⚡ GENERAL PROTECTION FROM ELECTRICAL INJURIES

or damage that could limit the tool from performing its intended function or could increase the potential for an incident. If the tool is damaged, discard the tool.

The manufacturer must mark voltage-rated hand tools with the following:

- Manufacturer's name or trademark
- Type or product reference (manufacturer's part number)
- Double triangle symbol **(see Figure 49)**
- AC voltage, such as 1,000 V
- Year of manufacture

ASTM F1505 *Requirements*

ASTM F1505 defines the testing of voltage-rated (insulated and insulating) hand tools.

- Unless specifically approved for use at low temperatures, the tool may be used in ordinary atmospheric temperatures ranging from 20°C to 70°C (68°F–158°F).

FIGURE 49 Double triangle symbol. (Courtesy of Charles R. Miller)

🔧 GENERAL PROTECTION FROM ELECTRICAL INJURIES

- The standard requires the mechanical integrity of the tools to meet the same requirements as nonrated hand tools.

- Only hand tools that meet the requirements of *ASTM F1505* or *IEC 60900* should be purchased.

- A storage toolbox could be purchased with the tools.

- The tools should be kept in the storage toolbox when not in use.

- The toolbox, with the tools, should be stored in a location that is clean and dry.

Temporary Protective Grounding Equipment

A de-energized electrical distribution circuit could be re-energized by several different means, which creates an unsafe condition.

- If the conductors are on cross arms, an energized conductor from a different circuit could fall onto one or more conductors of the de-energized circuit.

- If the circuit has equipment that is connected to multiple sources of energy, a second or third energy source could be operated to re-energize the de-energized circuit.

- The equipment could be back-fed through a transformer from a utilization circuit. In some instances, lockout/tagout could be in place and incorrectly implemented.

In each instance, a worker performing work on the distribution circuit could be electrocuted because of contact with the unintentionally energized conductor **(see Figure 50)**.

Where an exposure of this nature exists, the required electrically safe work condition does not exist. Workers must install temporary protective grounding equipment to control the potential exposure to shock and electrocution. When working on a de-energized conductor, the worker is likely to be a part of the de-energized circuit.

GENERAL PROTECTION FROM ELECTRICAL INJURIES

FIGURE 50   Overhead line work. (Courtesy of Win Henderson/FEMA)

- To minimize exposure, the worker must install a device that will keep the potential difference between the conductor and surrounding conductive objects to a minimum.

- Installing an adequately rated conductor from the circuit being worked on and adjacent grounded surfaces will provide low impedance and thereby low potential difference.

- Installing a grounding cluster ensures that the upstream overcurrent device will see a short circuit and operate in a short time mode.

When current flows through an electrical conductor, a magnetic field is produced in the vicinity of the conductor.

- The strength of the field depends on the amount of current flow, and the direction of the magnetic field depends on the direction of the current flow in the conductor.

⚡ GENERAL PROTECTION FROM ELECTRICAL INJURIES

- The magnetic field associated with each conductor interacts with other magnetic fields and some other metallic components that might be nearby.

- When the amount of current flow is in the range of the ampacity rating of the conductor, the strength of the associated magnetic field produces little effect.

- However, when the amount of current approaches the available short-circuit range of a distribution circuit, the magnetic fields can become excessive.

- When the magnetic fields surrounding each conductor of a three-phase circuit interact with each other, substantial physical forces result. The construction of the ground cluster must be capable of conducting the maximum available fault current in a circuit.

- As the amount of current increases, the strength of the magnetic lines of force also increases.

- When current is flowing in multiple conductors that are physically close to each other, the lines of magnetic force interact, resulting in strong physical forces applied to the conductors.

- If the conductors are a part of a ground cluster, the physical force tends to propel the conductor according to the interaction of the lines of magnetic force.

Performance Requirements

ASTM F855, Standard Specifications for Temporary Protective Grounds to Be Used on De-energized Electric Power Lines and Equipment, defines performance requirements for components of these ground clusters and for the complete assembly. Although the possibility exists for employees to assemble adequate protective grounds, the adequacy of the overall construction is not dependable without performing tests defined in the standard. Manufacturers perform tests as a routine part of the manufacturing process and assign fault duty ratings to the completed ground clusters. Only adequately rated ground clusters should be used **(see Figure 51)**; they are available through electrical distributors.

Temporary protective grounding equipment shall be placed at such locations and arranged in such a manner as to prevent each employee from being exposed to a shock hazard (i.e., hazardous differences in

FIGURE 51 Ground clusters with four conductors. (Courtesy of Salisbury Electrical Safety, LLC)

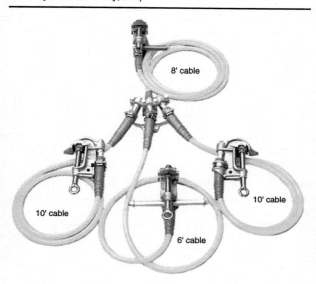

GENERAL PROTECTION FROM ELECTRICAL INJURIES

electrical potential). The location, sizing, and application of temporary protective grounding equipment shall be identified as part of the employer's job planning.

- Temporary protective grounding equipment shall be capable of conducting the maximum fault current that could flow at the point of grounding for the time necessary to clear the fault.

- Temporary protective grounding equipment and connections shall have an impedance low enough to cause immediate operation of protective devices in case of unintentional energizing of the electric conductors or circuit parts.

- Ground clusters must be marked to indicate the rating assigned by the manufacturer.

- After determining the necessary fault duty rating and selecting a ground cluster, the conductors, clamps, and connecting points should be visually inspected to ensure that the components of the ground cluster have not been damaged.

- If any sign of damage to the conductors or clamps is found, select another ground cluster.

- Installing temporary protective grounding equipment is one step required to establish an electrically safe work condition.

- Until the ground cluster has been satisfactorily installed, consider the conductors energized.

- Wearing appropriate arc-rated protection, install the ground cluster with a live-line tool or while wearing voltage-rated protective equipment.

- The first connection should be to an adequately sized grounding conductor. Subsequent connections should be made to each phase conductor.

GENERAL PROTECTION FROM ELECTRICAL INJURIES

- Ground clusters should be removed in reverse order. Remove the connection to the grounded conductor after all connections to a phase conductor are removed.

- OSHA requires a zone of equipotential be established for exposed overhead conductors. A zone of equipotential means that all metal components, including conductors, are grounded in such a way that the worker is unlikely to reach outside the zone of protection (see Figure 52).

FIGURE 52   Illustration of zone equipotential.

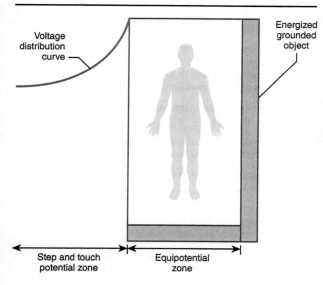

⛏ GENERAL PROTECTION FROM ELECTRICAL INJURIES

- If the overhead line receives a lightning discharge at another location, the safety grounds might protect the worker, regardless of the location of the strike.

- Ground clusters that have been repaired or modified must be tested to ensure that the repaired equipment will pass the standard 30-cycle or 15-cycle voltage-drop values permitted by *ASTM F855*.

- Ground clusters should be subjected to the 3-cycle or 15-cycle voltage-drop tests defined in the standard on a regular basis as determined by conditions of use. However, the test interval must not exceed 3 years.

⚡ FIRST AID

The first line of defense is turning off the power and establishing an electrically safe work condition. As previously discussed, turning off the power is only one step in establishing and verifying an electrically safe work condition. In addition, the best protection from the consequences of severe injury or illness is the knowledge of first aid practices and the ability to act in an emergency situation. The Emergency Care and Safety Institute is an educational organization created for the purpose of delivering the highest quality training to laypeople and professionals in the areas of first aid, CPR, AED, bloodborne pathogens, and related safety and health areas. The content of the training materials used by the Emergency Care and Safety Institute is approved by the American Academy of Orthopaedic Surgeons (AAOS) and the American College of Emergency Physicians (ACEP), two of the most respected names in injury, illness, and emergency medical care.

Visit www.ECSInstitute.org for more information.

Scene Size-Up

When approaching the scene of an emergency, take a few seconds to size up the scene to assess the following:

- Danger to the rescuer and to the victim. Scan the area for immediate dangers to yourself or to the victim. If the scene is unsafe, make it safe. If you are unable to do so, do not enter.

- Type of problem—injury or illness. This helps to identify what is wrong.

- Number of victims. Determine how many people are involved. There may be more than one person, so look around and ask about others.

If there are two or more victims, first check those who are not moving or talking. These are the individuals who may need your help first.

🔲 FIRST AID

How to Call for Help

To receive emergency assistance of every kind in most communities, simply call 9-1-1. At some government installations and industrial sites, an additional system may apply. This should be an element of a job safety plan and included in a job briefing. In any case, be prepared to tell the emergency medical services (EMS) dispatcher the following:

- Your name and phone number

- Exact location or address of emergency

- What happened

- Number of people

- Victim's condition and what is being done for the victim

Do not hang up until the dispatcher hangs up—the EMS dispatcher may be able to tell you how to care for the victim until the ambulance arrives.

Airway Obstruction

Management of Responsive Choking Victim

1. Check the victim for choking by asking, "Are you choking?" A choking person is unable to breathe, talk, cry, or cough.

2. Have someone call 9-1-1.

3. Position yourself behind the victim and locate the victim's navel.

4. Place a fist with the thumb side against the victim's abdomen just above the navel, grasp it with the other hand, and press it into the victim's abdomen with quick inward and upward thrusts. Continue thrusts until the object is removed or the victim becomes unresponsive.

If the victim becomes unresponsive, call 9-1-1 and give CPR.

🖐 FIRST AID

Adult Cardiopulmonary Resuscitation (CPR)

1. Check responsiveness by tapping the victim and asking, "Are you okay?"

2. At the same time, check for breathing by looking for chest rise and fall. If the victim is unresponsive and not breathing, he or she needs CPR (step 4).

3. Have someone call 9-1-1, and have someone else retrieve an AED if available.

4. Perform CPR.
   - Place the heel of one hand on the center of the chest between the nipples. Place the other hand on top of the first hand.
   - Depress the chest 2 inches.
   - Give 30 chest compressions at a rate of at least 100 compressions per minute, allowing the chest to return to its normal position after each compression.
   - Tilt the head back and lift the chin. Pinch the nose, and give 2 breaths (1 second each).

5. Continue cycles of 30 chest compressions and 2 breaths until an AED is available, the victim shows signs of life, EMS takes over, or you are too tired to continue.

Bleeding

1. For a shallow wound, wash it with soap and water, and flush with running water. Apply an antibiotic, and cover the wound with a clean dressing.

2. For a deep wound, do not attempt to clean the wound, but just stop the bleeding. Deep wounds require cleaning by a medically trained person. Cover large, gaping wounds with sterile gauze pads, and apply pressure to stop the bleeding. Secure the gauze pad snugly with a bandage.

⎏ FIRST AID

Protect yourself against diseases carried by blood by wearing disposable medical exam gloves, using several layers of cloth or gauze pads, using waterproof material such as plastic, or having the victim apply pressure using his or her own hand.

Burns

Care for Burns

1. Stop the burning! Use water or smother flames.

2. Cool the burn. Apply cool water or cool, wet cloths until pain decreases (usually within 10 minutes).

3. Apply aloe-vera gel on first-degree burns (skin turns red). Apply antibiotic ointment on second-degree burns (skin blisters). Apply nonstick dressing on second- and third-degree burns (full thickness; penetrates skin layers, muscle, and fat).

Seek medical attention if any of these conditions exist:

- Breathing difficulty

- Head, hands, feet, or genitals involved

- Victim is elderly or very young

- Electricity or chemical exposure involved

- Second-degree burns cover more than an area equivalent to the size of victim's entire back or chest

- Any third-degree burns

Electrical Burns

- Check the scene for electrical hazards.

- Check breathing; CPR may be needed.

🔌 FIRST AID

Frostbite

Recognizing Frostbite

The signs of frostbite include the following:

- White, waxy-looking skin

- Skin feels cold and numb (pain at first, followed by numbness)

- Blisters, which may appear after rewarming

Care for Frostbite

1. Move the victim to a warm place.

2. Remove any wet/cold clothing and any jewelry from the affected part.

3. Seek medical care.

Heart Attack

Recognizing a Heart Attack

Prompt medical care at the onset of a heart attack is vital to survival and the quality of recovery. This is sometimes easier said than done because many victims deny they are experiencing something as serious as a heart attack. The signs of a heart attack include the following:

- Chest pressure, squeezing, or pain lasting more than a few minutes. It may come and go.

- Pain spreading to either shoulder, the neck, the lower jaw, or either arm.

- Any or all of the following: weakness, dizziness, sweating, nausea, or shortness of breath.

🔲 FIRST AID

Care for a Heart Attack

1. Seek medical care by calling 9-1-1. Medications to dissolve a clot are available but must be given early.

2. Help the victim into the most comfortable resting position.

3. If the victim is alert, able to swallow, and not allergic to aspirin, give one adult aspirin or four chewable aspirin.

4. If the victim has been prescribed medication for heart disease, such as nitroglycerin, help the victim use it.

5. Monitor the victim's breathing.

Heat Cramps

Recognizing Heat Cramps

The signs of heat cramps include the following:

- Painful muscle spasms during or after physical activity

Care for Heat Cramps

1. Have the victim stop activity and rest in a cool area.

2. Stretch the cramped muscle.

3. If the victim is responsive and not nauseated, provide water or a commercial sport drink (such as Gatorade® or Powerade®).

Heat Exhaustion

Recognizing Heat Exhaustion

The signs of heat exhaustion can include the following:

- Heavy sweating

- Severe thirst

🔌 FIRST AID

- Weakness

- Headache

- Nausea and vomiting

Care for Heat Exhaustion

1. Have the victim stop activity and rest in a cool area.

2. Remove any excess or tight clothing.

3. If the victim is responsive and not nauseated, provide water or a commercial sport drink (such as Gatorade® or Powerade®).

4. Have the victim lie down.

5. Apply cool packs to the armpits and to the crease where the legs attach to the pelvis.

Seek medical care if the condition does not improve within 30 minutes.

Heatstroke

Recognizing Heatstroke

The signs of heatstroke can include the following:

- Extremely hot skin

- Dry skin (may be wet from strenuous work or exercise)

- Confusion

- Seizures

- Unresponsiveness

🔲 FIRST AID

Care for Heatstroke

1. Call 9-1-1.

2. Cool the victim immediately by whatever means possible: cool, wet towels or sheets to the head and body accompanied by fanning, and/or cold packs against the armpits, sides of neck, and groin.

3. If the victim is unresponsive and not breathing, begin CPR.

Hypothermia

Recognizing Hypothermia

The signs of hypothermia include the following:

- Uncontrollable shivering
- Confusion, sluggishness
- Cold skin (even under clothing)

Care for Hypothermia

1. Get the victim out of the cold.

2. Prevent heat loss by:
   - Replacing wet clothing with dry clothing
   - Covering the victim's head
   - Placing insulation (such as blankets, towels, coats) beneath and over the victim

3. Have the victim rest in a comfortable position.

4. If the victim is alert and able to swallow, give him or her warm, sugary beverages.

5. Seek medical care for severe hypothermia (rigid muscles, cold skin on abdomen, confusion, or lethargy).

🔌 FIRST AID

Ingested Poisons

Recognizing Ingested Poisoning

The signs of ingested poison include the following:

- Abdominal pain and cramping

- Nausea or vomiting

- Diarrhea

- Burns, odor, or stains around and in the mouth

- Drowsiness or unresponsiveness

- Poison container nearby

Care for Ingested Poison

1. Try to determine which poison was swallowed and how much.

2. For a responsive victim, call the poison control center for instructions: 800-222-1222. For an unresponsive victim, call 9-1-1.

3. Place the victim on their side if vomiting occurs.

Shock

Recognizing Shock

The signs of shock include the following:

- Altered mental status (agitation, anxiety, restlessness, and confusion)

- Pale, cold, and clammy skin, lips, and nail beds

- Nausea and vomiting

- Rapid breathing

- Unresponsiveness (when shock is severe)

🔲 FIRST AID

Care for Shock

Even if there are no signs of shock, you should still treat seriously injured or suddenly ill victims for shock.

1. Place the victim on their back.

2. Place blankets under and over the victim to keep them warm.

3. Call 9-1-1.

Stroke

Recognizing a Stroke

The signs of a stroke include the following:

- Sudden weakness or numbness of the face, an arm, or a leg on one side of the body

- Blurred or decreased vision, especially on one side of the visual field

- Problems speaking

- Dizziness or loss of balance

- Sudden, severe headache

- Sudden confusion

Care for a Stroke

1. Call 9-1-1.

2. If the victim is responsive, have them rest in the most comfortable position. This is often on their back with the head and shoulders slightly elevated.

3. If vomiting, roll the victim to their side (recovery position).

🔌 FIRST AID

First Aid, Rescue, and CPR

NFPA 70E requires employees who are exposed to shock hazards and those responsible for taking action in case of emergency to be trained in methods of safe release for victims in contact with exposed energized electrical conductors or circuit parts. Employees responsible for responding to medical emergencies shall also be instructed in methods of first aid and emergency procedures. Training shall occur at a frequency that satisfies the requirements of the certifying body. *NFPA 70E* also requires the employer to certify annually that the employee has been trained in approved methods of resuscitation, including CPR and automatic external defibrillator (AED) use. OSHA regulations require that at least one person on each job site be trained in first aid and CPR.

Many electrical contractors require foremen and general foremen to have first aid training. First aid, CPR, and AED training are available from several sources, including:

- Emergency Care and Safety Institute, www.ECSInstitute.org

- Local fire and rescue departments

- Community colleges

Many industrial plants, other facilities, and construction projects maintain a nurse on site.

🔋 REFERENCES

ANSI Z87.1, American National Standard for Occupational and Educational Personal Eye and Face Protection Devices. New York, NY: American National Standards Institute, 2015.

ANSI Z89.1, American National Standard for Industrial Head Protection. New York, NY: American National Standards Institute, 2014.

ANSI Z89.2, Safety Requirements for Industrial Protective Helmets for Electrical Workers. (Superseded by *ANSI Z89.1* and no longer published, although it remains a reference in *29 CFR 1910.6.*)

ANSI Z535, Series of Standards for Safety Signs and Tags. New York, NY: American National Standards Institute, 2017.

ASTM D120, Standard Specification for Rubber Insulating Gloves. Conshohocken, PA: American Society of Testing and Materials, 2014.

ASTM D1048, Standard Specification for Rubber Insulating Blankets. Conshohocken, PA: American Society of Testing and Materials, 2020.

ASTM D1051, Standard Specification for Rubber Insulating Sleeves. Conshohocken, PA: American Society of Testing and Materials, 2014.

ASTM D6413/D6413M, Standard Test Method for Flame Resistance of Textiles (Vertical Test). Conshohocken, PA: American Society of Testing and Materials, 2015.

ASTM F479, Specification for In-Service Care of Insulating Blankets. Conshohocken, PA: American Society of Testing and Materials, 2017.

ASTM F496, Specification for In-Service Care of Insulating Gloves and Sleeves. Conshohocken, PA: American Society of Testing and Materials, 2020.

🔌 REFERENCES

ASTM F711, Standard Specification for Fiberglass-Reinforced Plastic (FRP) Rod and Tube Used in Live Line Tools. Conshohocken, PA: American Society of Testing and Materials, 2017.

ASTM F855, Standard Specifications for Temporary Protective Grounds to Be Used on De-energized Electric Power Lines and Equipment. Conshohocken, PA: American Society of Testing and Materials, 2019.

ASTM F1116, Standard Test Method for Determining Dielectric Strength of Dielectric Footwear. Conshohocken, PA: American Society of Testing and Materials, 2014.

ASTM F1449, Standard Guide for Industrial Laundering Care and Maintenance of Flame Resistant or Arc Rated Clothing. Conshohocken, PA: American Society of Testing and Materials, 2020.

ASTM F1505, Standard Specification for Insulated and Insulating Hand Tools. Conshohocken, PA: American Society of Testing and Materials, 2016.

ASTM F1506, Standard Performance Specification for Flame Resistant and Electric Arc Rated Protective Clothing Worn by Workers Exposed to Flames and Electric Arcs. Conshohocken, PA: American Society of Testing and Materials, 2020.

ASTM F1959/F1959M, Standard Test Method for Determining the Arc Rating of Materials for Clothing. Conshohocken, PA: American Society of Testing and Materials, 2014.

ASTM F2413, Standard Specification for Performance Requirements for Protective (Safety) Toe Cap Footwear. Conshohocken, PA: American Society of Testing and Materials, 2018.

ASTM F2178, Standard Specification for Arc Rated Eye or Face Protective Products. Conshohocken, PA: American Society of Testing and Materials, 2020.

REFERENCES

Doughty, Richard L., et al. "Predicting Incident Energy to Better Manage the Electric Arc Hazard on 600V Power Distribution Systems." PCIC Paper PCIC-98-36. Paper presented at the Forty-Fifth Annual Conference of the IAS/IEEE Petroleum and Chemical Industry Committee, Indianapolis, IN, September 28–30, 1998.

Doughty, Richard L., et al., "Testing Update on Protective Clothing and Equipment for Electric Arc Exposure." PCIC Paper PCIC-97-35. Paper presented at the Forty-Fourth Annual Conference of the IAS/IEEE Petroleum and Chemical Industry Committee, Banff, Alberta, September 15–17, 1997.

IEC 60900, Live Working—Hand Tools for Use up to 1000 V.a.c. and 1500 V.d.c. Geneva, Switzerland: International Electrotechnical Commission, 2019.

IEEE Standard 1584, IEEE Guide for Performing Arc-Flash Hazard Calculations. New York, NY: Institute of Electrical and Electronics Engineers, 2018.

Lee, Ralph H. "Pressures Developed by Arcs." *IEEE Transactions on Industry Applications*, Vol. 1A-23, No. 4, July/August 1987. Piscataway, NJ: IEEE, 1987.

The National Electrical Safety Code (ANSI/IEEE C2). New York, NY: Institute of Electrical and Electronics Engineers, 2017.

NFPA 70®, *National Electrical Code*® (NEC®). Quincy, MA: National Fire Protection Association, 2020.

NFPA 70E®, *Standard for Electrical Safety in the Workplace*®. Quincy, MA: National Fire Protection Association, 2021.

Phillips, Jim P. E., *Complete Guide to Arc Flash Hazard Calculation Studies*. Scottsdale, AZ: Brainfiller, 2011.

⧉ REFERENCES

UL 1244, Standard for Electrical and Electronic Measuring and Testing Equipment. Northbrook, IL: Underwriters Laboratories, 2000.

U.S. Department of Labor. Occupational Safety and Health Administration. OSHA Regulations 29 CFR 1910.132-139, Subpart I, "Personal Protective Equipment." Washington, DC.

U.S. Department of Labor. Occupational Safety and Health Administration. OSHA Regulations 29 CFR 1910.137, "Electrical Protective Equipment." Washington, DC.

U.S. Department of Labor. Occupational Safety and Health Administration. OSHA Regulations 29 CFR 1910.269, "Electric Power Generation, Transmission, and Distribution." Washington, DC.

U.S. Department of Labor. Occupational Safety and Health Administration. OSHA Regulations 29 CFR 1910.300-399, Subpart S, "Electrical." Washington, DC.